Jerry Baker's

ULTIMATE
HOUSEHOLD
TONICS
BOOK!

www.jerrybaker.com

Jerry Baker's

ULTIMATE
HOUSEHOLD
TONICS
BOOK!

476 *Fantastic Formulas*

to keep your kitchen, laundry, floors,

pets, furniture, and houseplants

in tip-top shape!

by Jerry Baker

Published by American Master Products, Inc.

Executive Editor: Kim Adam Gasior
Managing Editor: Cheryl Winters-Tetreau
Research and Writer: Zach Gasior
Copy Editor: Nanette Bendyna
Production Editor: Debby Duvall
Interior Design and Layout: Sandy Freeman
Indexer: Nan Badgett

Publisher's Cataloging-in-Publication
(Provided by Quality Books, Inc.)

Baker, Jerry.
     Ultimate household tonics book : 476 fantastic
  formulas to keep your kitchen, laundry, floors, pets,
  furniture, and houseplants in tip-top shape! / Jerry Baker.
     p. cm.
     Includes index.
     ISBN-13: 978-0-922433-02-5
     ISBN-10: 0-922433-02-X

    1. Home economics.   I. Title.

TX158.B35 2012   640
                     QBI11-600207

Printed in the United States of America
4 6 8 10 9 7 5 3 hardcover

CONTENTS

CHAPTER 2

BATHROOM
Brighteners

CHAPTER 3

LAUNDRY
Lowdown

CHAPTER 4

ROOM for Improvement

CHAPTER 5

FURNITURE
Fixers

CHAPTER 6

FLOORS
a Go-Go

CHAPTER 7

Super STAIN Removers

CHAPTER 8

Happy HOLIDAYS

CHAPTER 9

PET
Projects

CHAPTER 10

BIRD
Buffet

CHAPTER 11

PEST
Peeves

CHAPTER 12

HOUSEPLANT
Hoedown

CHAPTER 13

OUTDOOR
Chores

INTRODUCTION

IF YOU'RE LIKE MOST FOLKS, your cupboards are crammed full of cleansers, removers, polishes, and so many powders, liquids, and gels that you forget why you even bought them in the first place! If this sounds about right, then I'm here to help you clear out the commercial concoctions and replace them with good old-fashioned DIY tonics that'll liven up your home—from your front porch to your back deck, and every teapot, tile, toilet, quilt, cabinet, couch, pooch, finch, and fern in between. Whew!

The first step to keeping any house in tip-top shape is to tackle the cleaning issues. You'll learn how to bewitch your kitchen with stainless steel solutions and detergent-free dish washers, brighten your bathrooms with "green" scouring powders and citrus-sweet formulas, zap stubborn stains with tough-as-nails tonics, and gently rinse delicate duds with simple, pure potions. Plus, your floors will look great with my nontoxic carpet cleaners and super-duper spiffer-uppers for vinyl, wood, tile, and more.

But that's not all—my Ultimate Household Tonics go well beyond just cleaning and scrubbing. You'll also be able to fix furniture flaws fast using a splash of "Salad Dressing" Scratch Remover, reward your furry friends with a batch of Just for Love Dog Delights, keep your houseplants happy and healthy with my famous Fabulous Foliage Food, and show household bugs who's boss with a heaping helping of Beat the Ant Blues Batter. And when the holidays roll around, you'll sail through stress-free with oh-so-easy homemade soaps and other

great gifts, and still have plenty of time to get crafty with the kids.

Moving outdoors, you'll be ready to rock with a bucketful of handy helpers. I've included bunches of bird-pleasing bounties, including my Flicker Flapdoodle, patio-perfecting potions like my Crackerjack Concrete Cleaner, and a Longer-Life Battery Bath that's sure to keep the old jalopy purring like a kitten.

Throughout this book, you'll also find loads of fascinating features to help unleash the power-packed potential of my Ultimate Tonics. For instance, **PINCHING PENNIES** shares my money-saving advice on everything from

deodorizing a dishwasher to trapping pesky pests to removing rust from your patio furniture.

THAT'S BRILLIANT! offers tips that are both practical and surprising—like sprucing up grungy sneakers with carpet shampoo, tidying up your oven's interior with tea leaves, and keeping bugs off your freshly painted house with citronella oil. Don't be surprised if you end up saying, *"Why didn't I think of that?"*

And when you need to get something done fast, check out the **SUPER SHORTCUTS**. These terrific tricks tackle your toughest issues in no time at all. For instance, you'll learn how to make ink stains disappear from leather with milk, soak away burned-on food in pots and pans like magic, and use lemon juice for almost effortless every day, every way cleaning.

I also reached deep into my mailbag to come up with the best **FAQ** (Frequently Asked Questions) I could find, featuring my easy-to-apply answers to your most perplexing problems. Whether you're wrestling with a clogged bathroom drain, need to know the best way to zap backyard bugs, or are struggling to remove old bumper stickers from your car, here's where I let you in on my best-kept secrets.

Finally, I'd never gather up this much information and leave out my Grandma Putt's advice. After all, she taught me almost everything I know about taking care of the old home-stead. So now it's time to pass that lore along to you. **GRANDMA'S OLD-TIME TIPS** are exactly that—words of wisdom that'll help you in dozens of different ways, including how to brighten up wicker, polish wood, and keep your houseplants humming a happy tune.

So roll up your sleeves and dive right in. Mix up my Ultimate Household Tonics and you'll soon be on your way to a cleaner, "greener" house; happier, healthier living; and—most important of all—a calmer, more contented you!

CHAPTER 1

Bewitchin' in the
KITCHEN

No matter what's bewitchin' in your kitchen, cleanup always seems to be a chore. But it doesn't have to be. Check out these terrific tonics for cabinets, counters, cooktops, and more. Most are made with ingredients you already have around the house, which means you'll be able to mix 'em up and put 'em to good use spraying, swiping, and sanitizing. Before you know it, your kitchen will be fresh as a daisy and rarin' to go. So let's get cookin'!

All-Around Hard-Hitting Kitchen Cleaner

⅓ cup of rubbing alcohol
1 tsp. of ammonia
1 tsp. of dishwashing liquid
½ tsp. of lemon juice
3 cups of water

If you're looking for one cleaner that you can use over and over again on any possible mess that might occur in your kitchen, then this tough-as-nails formula is for you. It'll handle anything you throw at it—including the kitchen sink!

DIRECTIONS: Mix the ingredients, then pour the solution into a handheld sprayer bottle. Spray it on countertops, appliances, tile backsplashes, and even painted walls. Wipe the surface clean with a damp cloth or sponge.

"Just Baked" Aroma 🔺 THERE'S NOTHING quite as welcoming as a kitchen that smells like a wonderful treat just came out of the oven. But you don't have to bake a cake to send that enticing aroma wafting from your kitchen. Simply turn the oven on low and set a dish of vanilla extract in it. Your guests will think something is baking (and the secret scented "candle" will be nowhere in sight).

Now You're Cookin'! 🔺 IF YOU LOVE TO COOK, but hate the lingering odors, keep a small bowl of vinegar on top of your stove. It'll absorb heavy cooking odors—even from the strongest-smelling foods like fish and onions. Not so keen on vinegar? Put cinnamon and cloves in a saucepan with water and bring it to a boil. Let it simmer on the stove until the air is fresh.

PINCHING PENNIES

Are you looking for a simple (and economical) way to disinfect and deodorize the surfaces in your kitchen, bathroom, or baby's room? Well, look no further! Mix ½ cup of borax in 1 gallon of hot water, and go to town. You can either pour the solution into a handheld sprayer bottle, or mix it up in a bucket, dip a sponge in, and wipe dirt and germs away—it's your call.

Amazing Porcelain Stove Potion

This potion is so darn good, it'll spiff up your entire porcelain stove—including the oven! This all-purpose spray loosens up crud and cuts through greasy spills like nobody's business.

> 2 tsp. of white vinegar
> 1 tsp. of borax
> ½ tsp. of baking soda
> 1 squirt of grease-cutting dishwashing liquid
> 2 cups of water

DIRECTIONS: Measure the ingredients into a handheld sprayer bottle, and shake it well. Be sure the surface is cool, and then spray and wipe porcelain stove tops, drip pans, gas burner grates, ovens, and even the glass windows in oven doors. For dried-on spills, soak the spots with the potion, and wait about 15 minutes before wiping it up. Shake the bottle frequently as you work, and store it in a cool cabinet when you're finished. To avoid scratches, don't use on stainless steel or smooth-surface cooktops.

Soak 'n' Scrub 🔺 TO LOOSEN CRUSTY CRUD on broiler pans, drip pans, or burner rings, soak them in a sink full of hot, soapy water (use grease-cutting dishwashing liquid) with ¼ cup of ammonia added to it. Let them soak for at least two hours, or overnight, then finish the job with a nonabrasive plastic scrubbie.

A Tangy Solution 🔺 TO CLEAN UP BAKED-ON spills from any kind of stove top (or oven), mix lemon juice, water, and baking soda to make a paste. Cover the trouble spots, wait 15 minutes, and wipe the marks away.

That's Brilliant!

When you see yellow flames instead of blue ones around the rings that fire the gas burners on your stove—or no flame at all in some spots—don't rush to call a repairman. It just means that the holes around the rings are clogged with grease or dirt. To fire things back up, unclog the holes by carefully poking each one with a pipe cleaner or piece of fine wire, like a straightened paper clip. Don't use a toothpick because it can easily break off in the hole, and then you've got real trouble.

Antibacterial Cutting Board Wipe

> 2 parts water
> 1 part white vinegar
> Dishwashing liquid

This simple potion puts the power of plain old vinegar to work for you, killing common food bacteria. It's the acetic acid in the vinegar that works like magic—why, it makes even big, bad germs like salmonella beg for mercy!

DIRECTIONS: Mix the water and vinegar in a handheld sprayer bottle. Wash and rinse the cutting boards with dishwashing liquid and hot water, then wipe down both sides with the vinegar solution. If your cutting board is made of wood, let the solution sit for about five minutes before you dry the board, so that the germ-fighting vinegar can penetrate beyond the surface. There's no need to rinse because the vinegar smell will disappear as it dries!

Peroxide Wipe 🏠 HERE'S A QUICK AND EASY way to destroy germs on your cutting boards. First, wipe down the boards—both sides!—with my Antibacterial Cutting Board Wipe (above) and then follow it up with a second swipe of hydrogen peroxide for extra insurance. Simply pour some peroxide onto a paper towel, and rub it over the boards.

Banish the Smell 🏠 CUTTING BOARD SMELLING a little fishy? Or maybe the aroma came from garlic or onions. Whatever caused the scent, you can make it vamoose by wiping the board with lemon juice and rinsing it with cool water.

Q *My best plastic cutting boards are becoming grimy. It's getting to the point where food is remaining stuck in the nicks, even after I wash the boards. I got some new ones, but I don't want to run into the same problem again later. So when should I replace them?*

A Nicks and scratches give salmonella and other bacteria a great place to hide, and washing by hand isn't enough to clean all the germs out of those deep gouges in a plastic cutting board. So sanitize a battle-scarred plastic board in the dishwasher, where the heat will help kill the germs. And then replace the board when it gets very badly scratched.

Baking Soda Oven Scrape

Baking soda
White vinegar
Water

Does this sound familiar? You've prepped, peeled, and rolled to your heart's content, and now your apple pie is in the oven. Then, big-time trouble hits when the filling bubbles over and dribbles onto the bottom of the oven. It not only fills the house with the smell of burnt baked goods, it also creates a thick glob of gunk for you to clean up later. Stop worrying and reach for the baking soda. It'll clean the mess while you sleep!

DIRECTIONS: Before you go to bed, sprinkle a thick layer of baking soda on the bottom of the cool oven. Combine equal parts of vinegar and water in a handheld sprayer bottle, and spray the soda to wet it down, close the oven door, and say good night. By morning, all of the grunge will have loosened up, and you can wipe it out with a wet sponge.

Vinegar Wipe ✂ A REGULAR WIPE-DOWN with white vinegar will go a long way toward keeping an oven spic-and-span. Vinegar cuts through greasy crud quick as a wink, so just pour some on a sponge, wipe down the oven, rinse it with a moist sponge, and you're good to go!

Flow Down ✂ A BAKING SODA SPRAY will make the blackened gunk on oven walls give up its grip and slide down to the bottom. Put about 3 tablespoons of baking soda into a handheld sprayer bottle, fill the bottle with hot water, and shake it vigorously. Make sure the oven is cool, and spritz the walls with the mixture, shaking it frequently as you go. When the grime collects on the bottom, wipe it away.

Super Shortcuts

✂ To minimize your oven-cleaning chores, line a cookie sheet with foil and put it on the bottom rack underneath the dish you're baking in order to catch the spills. When you're done baking, just crumple up the dirty foil and toss it in the trash. If any spills happen to miss the cookie sheet and land on the oven, dip the crumpled foil into a paste made of baking soda and water, and use the foil as a scrubber to remove the goo before you trash it.

Burned-On Grease Remover

> 1 part cream of tartar
> 1 part white vinegar

Tired of scrubbing away at your enamel broiler pan to get the burned-on grease out? This laborsaving potion works like magic and lets the crud know it's time to hightail it on outta there.

DIRECTIONS: Mix the cream of tartar and vinegar in a small bowl; you'll need several tablespoons of each for most jobs. Using a cloth or cotton ball, apply the mixture to the gunked-up spots on the edges or sides of your pans. Pour the rest of the mixture into the bottom of the broiler pan, tilting the pan so that it spreads out. Let the mixture sit for about 10 minutes, then go over the pan with a nonabrasive plastic scrubbie to lift off the crud. If any stubborn spots remain, repeat the treatment. Then wash your pan in warm, soapy water, let it air-dry, and put it back in place.

Baking Soda Scrub 🏠 TO SCRUB AWAY STAINS from smooth-surfaced enamel like your cookware or bathtub, make a paste of baking soda and a little water. It'll rub the spots away without scratching the surface. For more scrubbing action, use a nonabrasive plastic scrubbie instead of a cloth to apply the paste. After you wipe the baking soda off, finish the job with a vinegar rinse.

Go Along, Grease 🏠 GOT BURNED-ON GREASE in your enamel pans? Fill 'em with water, and drop in six Alka-Seltzer® tablets. Let the pans soak for an hour, and wash as usual. (This ploy works just as well with non-enamel pots and pans.)

PINCHING PENNIES

Don't worry about going out and finding just the right cleaning product. Stick to good old (cheap) soap and water when you clean decorative enamelware like plates, jewelry, and picture frames, so you don't damage any paint that may be on the surface. Dust the piece first to remove loose dirt, and then give it a gentle sponge bath in the sink. Pad the bottom of the sink with a rubber mat beforehand, just in case you develop a sudden case of butterfingers!

Copper Shine Solution

> 1 cup of white vinegar
> 1 tbsp. of salt
> Water

Copper can get dingy over time, just like everything else. So you have to tread lightly when you clean it, because scouring powders and other abrasives can leave fine scratches in copper. If you prefer a nice sleek shine, try this trick instead.

DIRECTIONS: Put your copper item in a large pot, cover it with water, and add the vinegar and salt. Put the pot on the stove and bring the water to a boil. Then turn down the heat, and let it simmer for a few hours so the acid can go to work on that tarnish. Let it cool, wash your copper piece in soapy water, dry it thoroughly, and you're done.

Lemon Copper Cleaner 🏠 KEEP COPPER POTS and pans sparkling bright by rubbing them with half a lemon dipped in salt. Then wash them in warm, soapy water and rinse thoroughly. (Any trace of the fruit's acid on the metal can cause pitting.)

Bring Back the Luster 🏠 SOME PEOPLE LIKE a little verdigris on their brass and copper pieces. But if you're not one of those folks—or if the metal has simply turned too green for your taste—bring back the original luster by rubbing the surface with a half-and-half solution of ammonia and salt, and rinse with clear water.

GRANDMA'S OLD-TIME TIPS

It's fairly common to have a set of pots and pans with a thin coating of copper on the bottoms. It's also common for that copper to get tarnished over time. Here's how Grandma Putt kept her copper-bottomed pans clean and shiny. She mixed up some lemon juice and cream of tartar to make a thick paste. Then she rubbed it onto the copper with her fingers, and let it sit for about 10 minutes. When she wiped the paste off and washed the bottom, it was as shiny and bright as a full moon. A quick buff with a clean rag added the final, finishing touch.

Countertop Crud Buster

We eat on them, we prepare meals on them, sometimes we even sit on them. Whatever you use your countertops for, they're bound to build up dirt and grime over time. This simple formula is just the ticket for tackling gunked-up kitchen counters.

> ½ cup of dishwashing liquid
> ¼ cup of baking soda

DIRECTIONS: Mix the ingredients in a small plastic bowl, then wipe the paste on your countertops with a damp sponge. Wring out the sponge, and wipe away the paste. Then towel-dry the countertop and watch it shine.

Lemon Stain Lifter ☛ TO GENTLY BLEACH stains out of a laminate countertop, rub a slice of lemon on the stain, and then let the lemon sit right on the dirty spot for about 15 minutes. Rinse the area clean with a damp cloth, and wipe it dry. If you still see a shadow of the stain, go ahead and repeat the treatment until it's gone.

Go, Gunk, Go ☛ GOT SOME DRIED-ON cookie dough, melted cheese, or other crusted gunk on your laminate countertop? Swipe it all away with this all-purpose spray: Fill a 1-quart handheld sprayer bottle with equal parts dishwashing liquid, bleach, and pine oil. Take aim and saturate the mess. Let it sit for a minute, then wipe off the counter with a clean, damp sponge or cloth.

Bleach Butcher's Block ☛ TO REMOVE STAINS from wooden cutting boards or butcher block countertops, soak a white dishcloth in undiluted bleach, and lay it over the marks. Let the cloth sit for 10 to 15 minutes, then rinse the area.

That's Brilliant!

Clean dull, dingy grout on ceramic tile countertops or backsplashes the easy way—with an old toothbrush. Just dip the brush into a solution of 1 teaspoon of bleach in 1 cup of water, and scrub your grout lines back to their pearly white beginnings.

Crackerjack Castile Dishwashing Liquid

> Liquid castile soap
> 8 drops of lemongrass oil
> 4 drops of bergamot oil
> 4 drops of rose oil

So what's the deal with castile? It's an all-natural cleanser, derived from vegetable oil, not animal fats or synthetics like most commercial dishwashing liquids. Castile isn't a brand, it's simply a type of soap, and it comes in liquid and bar forms. Use the liquid kind to whip up your very own dishwashing liquid. It's gentle on your hands but tough as nails when it comes to tackling the dirtiest dishes and gunkiest grime.

DIRECTIONS: Follow the directions on the bottle to dilute the castile soap. Pour 1 cup of the diluted soap into a clean reusable squirt bottle, then mix in the essential oils. Add 1 to 2 tablespoons of the solution to hot water to hand wash a sink full of dishes. It'll have all the power of those commercial brands, but at just a fraction of the price.

Grease, Be Gone! 🏠 WHEN YOUR DISHWASHER'S racks or walls develop a greasy coating, don't get down on your hands and knees to clean it up—grab a box of baking soda, and let the machine do the work itself. Pour 1 cup of baking soda into the dishwasher, and run it through the rinse cycle. It'll combine with the grease and turn it into soap, then go right down the drain, leaving your machine squeaky clean—and saving you wear and tear on your knees.

PINCHING PENNIES

Food residue, moldy gunk, and greasy buildup can make your dishwasher smell downright funky. So cut the crud with a simple bleach treatment, and save your wallet by staying away from the expensive cleaners. Pour 1/2 cup of bleach into the bottom of your empty dishwasher, and run it through a full cycle to sanitize and deodorize it.

Disposal Scrub-a-Dub

> 2 cups of ice cubes
> 1 cup of white or cider vinegar

Hard food waste like peach pits and bones are obviously big no-nos in the garbage disposal, but ice cubes are just the ticket for scrubbing a sour one down. With this potion, the blades fling the chunks against the drainpipe, where they help free up the gunk in the disposal, while the vinegar washes the stinky bacteria down the drain.

DIRECTIONS: With the garbage disposal turned off, put the ice cubes in the drain and pour in the vinegar. Turn on the cold water and run the disposal. Keep it going for at least 30 seconds after the clunking and thunking changes to a steady, high-pitched whine. Just don't try this trick when company's coming because it'll take the sharp smell of the vinegar an hour or two to disappear.

Cold Water Only ⬆ ALWAYS USE COLD WATER to wash garbage down a disposal. Hot water will melt the fats in the food, leaving a layer of grease in the pipes that can soon turn rancid and start to smell.

Give It a Java Jolt ⬆ SOME FOLKS STILL believe the old wives' tale that you should never put coffee grounds down a garbage disposal. Well, it just ain't true. In fact, coffee grounds actually help keep the blades sharp and the innards turning freely. So, by all means, give your disposal healthy helpings!

Q *There's a bit of an odor coming from my sink, but it's clean as a whistle. However, I noticed that the rubber guard in my disposal was looking less than spanking clean. Is there any way I can freshen it up?*

A The rubber cover that keeps food from splattering does get mighty slimy. So get rid of the goo in a hurry by scrubbing it with an old toothbrush dipped in hot, soapy water. The job is easy if you remove the cover, but if yours is permanently attached, lift the flaps to scrub the undersides. And make sure no one is anywhere near the switch while you're doing the job!

Easy Oven Cleaner

Use this powerful paste to wipe up occasional spills, or whenever your oven needs a good, thorough cleaning. Just be sure the oven is cool before you start to work.

> 2 parts water
> 1 part ammonia
> 1 part white vinegar
> Baking soda

DIRECTIONS: Combine the liquid ingredients in a large glass or ceramic bowl, and stir in enough baking soda to make a creamy paste. Rub the paste all over the bottom and sides of your oven, as well as the door and the inside of the window. Let the paste sit for about 30 minutes, then wipe it off. If any stubborn spots remain, use a nonabrasive plastic scrubbie and another dab of paste to cut through the crud. Rinse off the residue with a moist sponge, and your oven will be sparkling clean. To make it even easier for the paste to do its stuff, wipe any loose grime out of your oven before you apply it.

Hands-Off Oven Cleaning 🏠 DO YOU HAVE AN OVEN that's not self-cleaning? To do the job yourself the easy way, combine ¼ cup of ammonia and 2 cups of water in a glass baking dish. Put the dish in the oven, shut the door, and leave it overnight. In the morning, the grime will wipe away easily with a sponge.

Salt Those Spills 🏠 WHEN A STICKY PIE bubbles over onto the bottom of an oven, sprinkle the goo with a layer of salt while the oven is still hot. The salt will soak up the mess and keep the oven from smoking. As soon as the oven is cool, scrape off the hardened residue with a non-scratch spatula.

That's Brilliant!

When you combine the grease-cutting power of lemon juice and the abrasive action of salt, you have a great spot cleaner that'll scrub away spills in a flash—and make your kitchen smell good, too! Just mix 1 part fresh or bottled lemon juice with 1 part salt and apply it to the spills with a damp sponge. Let it sit for about 15 minutes so the juice can penetrate, then rub the spots with a nonabrasive plastic scrubbie. Wipe out the residue, and you're done!

Everyday Counter Cleaner

This mixture is just the ticket for removing dirt from laminate, acrylic, ceramic tile, or cultured marble countertops. Just don't use it on marble, granite, or other natural stone or the mild acids in the mixture could damage it.

> 1 cup of ammonia
> ½ cup of white vinegar
> ¼ cup of baking soda
> 1 gal. of water

DIRECTIONS: Mix the ingredients in a bucket, and sponge the solution onto your countertops to wipe away grime. Rinse with fresh water, and dry with a soft cloth to reveal the shine.

Long Life for Laminate 🏠 CLEANING STAINS OFF laminate countertops couldn't be simpler: Just use a nonabrasive plastic scrubbie to get rid of dried-on gunk. For stubborn stains like grape juice, make a paste of 3 parts baking soda to 1 part water, cover the stain, and let sit for 5 to 10 minutes before you wipe it away.

Fill It In 🏠 IF YOUR LAMINATE COUNTERTOP gets scratched or dinged, you can fill in those dirt-collecting gouges with a scratch-repair paste from your local hardware store. The paste comes in a variety of colors; if yours isn't available, you can blend two or more colors to match. Clean the damaged area with rubbing alcohol, fill the chip or scratch in with the repair paste, smooth it out, and scrape off any excess with a putty knife. Let the paste cure for 24 hours, and your counter will be as good as new!

GRANDMA'S OLD-TIME TIPS

Years ago, a hefty slab of slate was the work surface of choice in country kitchens like my Grandma Putt's. Nowadays, slate is back again in a big way with lots of special products made to care for it. But you don't have to spend your hard-earned cash—just use one of my Grandma's old tricks! Whenever her slate looked dull, she rubbed in a dab of mineral oil to give it a natural sheen that also prevented grease and other stains from being absorbed into the stone.

Fresh 'n' Clean Formica® Formula

½ cup of white vinegar
1 tsp. of dishwashing liquid
¼ tsp. of olive oil
½ cup of hot water

Formica may appear to be tough as nails, but it's actually a fairly fragile surface that can scratch easily and is quickly dulled by abrasive cleaners. So treat it right with this gentle formula.

DIRECTIONS: Pour all of the ingredients into a handheld sprayer bottle and swirl the solution around to mix well. Spray your countertops thoroughly and let the cleaner sit for a minute or two before wiping it away with a damp sponge. Rinse out the sponge with warm water between wipes, then sponge down the counters with more warm water to clean away any residue.

Fixes for Faux Stone NOT-QUITE-NATURAL STONE like Silestone®, which is a composite of crushed quartz and polymer resin, requires much less maintenance than its natural look-alikes because it won't soak up grease or other stains. You'll find this surface material super easy to clean, too—just give it a swipe with warm, soapy water, and wipe it dry.

Ants Away TO KEEP ANTS OFF your countertops, wipe the surfaces with a wet, squeezed-out paper towel. Then wipe again with full-strength white vinegar, and  rinse thoroughly. The vinegar will help remove the scent trails the ants laid down, so that others in the colony can't follow them to the food. Just make sure you take care of the source of the problem, and you'll keep your entire kitchen ant-free.

That's Brilliant!

Countertops can be made of tile, glass, marble, and even bronze. Basic cleaning is the same, though, no matter what the material. Wipe your counters clean daily with a damp cloth, and dry them thoroughly. That'll keep grunge under control. Then when it's time for a full-scale cleaning, refer to my tips for your individual counter's material.

Garbage Can Clean-'Em-Up

When your kitchen garbage can starts smelling like, well, garbage, take action ASAP because it'll only get worse. Start with this terrific tonic.

½ cup of white vinegar
2 tbsp. of dishwashing liquid
2 cups of hot water
Borax

DIRECTIONS: Mix the first three ingredients in the empty garbage can, then swish them around and scrub the bottom and the sides with a solution-soaked sponge. Rinse the can well and let it dry, then sprinkle borax across the bottom before replacing the bag. Sprinkle a new layer of borax into the can every month or so to keep the stink at bay.

How'd You Expect Garbage to Smell?

I KNOW, GARBAGE CANS aren't supposed to smell good, but that doesn't mean they have to smell bad! If you empty your can frequently, it shouldn't build up too much of an odor. To help prevent stench, sprinkle the bottom of your trash can with baking soda (borax will do, too) before replacing the plastic liner. It absorbs most bad odors, and you can use it to scour the can the next time you wash it.

Mmmmmm . . . Vanilla!

PLACE A FEW DROPS of vanilla extract on two or three cotton balls and drop them into your trash can. Vanilla has a powerful aroma that can mask almost anything—in an oh-so-delicious way!

Super Shortcuts

Industrial absorbent (the clay-based kind, available at home-improvement centers) can make your outdoor garbage can less of a "nosesore." Sprinkle a half inch or so of the material on the bottom of the can to absorb odor-causing grease and moisture. This will be especially helpful if you keep your garbage cans in the garage—the area won't retain any odors that happen to build up due to your stinky trash.

Garbage Disposal Deodorizer

½ of an orange
¼ cup of baking soda

Garbage disposals work hard, but sometimes they're responsible for stinkin' the whole kitchen up. Clear the air with this easy solution.

DIRECTIONS: Drop the ingredients into the opening, then turn on the cold water and run the disposal for 30 seconds. The citrus oil in the orange skin cuts any lingering (and smelly) grease in the disposal and lends a nice clean scent to the surrounding area, while the baking soda scrubs away built-up crud.

Soapy Flush 🏠 IF FOUL ODORS have started to seep out of your disposal, then send a sink full of soapy water down there to flush out any rotting slurry that's lingering in the pipes. Simply fill your sink about two-thirds of the way up with hot, soapy water, unplug the stopper, and switch on the disposal. The whirlpool of water will drive those smells right down the drain. Follow up by rinsing everything with cold water.

No Shortcuts, Please! 🏠 BAD SMELLS are the bane of garbage disposals, but it takes only a few seconds to keep the odors from building up. Instead of switching the disposal off as soon as you hear that high-pitched whine, let the water and disposal run for an additional 30 seconds. This trick will wash away the food residue that's splashed on the insides of the disposal and drainpipe, so the gunk doesn't build up and start to decompose.

PINCHING PENNIES

Save yourself some major repair bills—don't use your disposal to grind up any garbage that's full of stringy fibers. Toss corn husks, pea pods, string beans, celery, and artichoke leaves onto your compost pile instead, so that the fibers don't tangle up the disposal blades. If you've already sent the stuff down the drain and the blades have ground to a halt, *turn off the circuit breaker* to make sure you stay safe, and pull out as much of the stuff as you can with a pair of pliers (not your hands!).

Granite Countertop Shine Booster

¼ cup of rubbing alcohol
1 drop of dishwashing liquid
3 cups of water

Granite countertops look great in the kitchen. They give the entire place a sophisticated look—but only if they're spotless. If your granite countertops are looking dull as dishwater, try this magic mixture to wipe the film away and leave 'em fresh and brand-spankin' clean.

DIRECTIONS: Pour the ingredients into a handheld sprayer bottle, swishing to combine. Spritz the cleaner over the countertops, and then wipe dry with a microfiber cleaning cloth. It'll take only a minute to restore the shine—and there's no need to rinse!

Clean as You Cook 🏠 STAINS ARE NO FUN to scrub away, no matter what kind of countertops you have, which is why it's a great idea to wipe up all spills and messes as soon as they occur. This handy habit pays off big-time on some of the newer countertop materials because grease, wine, and other stains can quickly ruin the look of granite and other elegant surfaces. So whether you're cooking up a fancy dinner, or merely fixing yourself a snack, remember to wipe as you go, and you'll make stains a thing of the past!

Polish with Wax 🏠 CAR WAX was my Grandma Putt's favorite polish for her Formica® countertops and kitchen table. She rubbed it on with a soft cloth and buffed the surface with a second cloth. Why, by the time she was done, I could practically see myself in the counters and tabletops. They were that shiny!

Super Shortcuts

Save yourself a lot of time cleaning your granite countertops. Sleek finishes, such as shiny granite or satiny stainless steel, can show every little water spot after you wipe up spills, unless you dry them thoroughly. So grab a paper towel or a microfiber cloth, and use long, lengthwise strokes to leave a spot-free surface behind.

Herb Garden Scouring Powder

This all-natural, mild abrasive is perfect for cleaning your splotchy, grungy kitchen sink. It's tough enough to handle the grime, but it'll leave behind a gentle aroma you can't get from harsh chemical cleaners.

DIRECTIONS: Mix all of the ingredients, then place them in a container with a plastic lid. Poke several holes through the lid with an ice pick or screwdriver, and the cleaner is ready to be put to use. Just shake the powder sparingly into the sink and use a damp sponge to scrub the surface. Rinse away the residue with warm water.

> 2 cups of baking soda
>
> 2 tbsp. of crushed rosemary leaves
>
> 2 tbsp. of ground sage leaves
>
> 1 tbsp. of cream of tartar

Porcelain Picker-Upper HERE'S A SUPER-SIMPLE way to clean a porcelain sink—even if the dirt has been building up for a while. Cover the surface with a thick layer of paper towels, then saturate them with a half-and-half solution of bleach and water. Wait about five minutes, remove the towels (wearing rubber gloves), and rinse with clear water.

GRANDMA'S OLD-TIME TIPS

Always keep Grandma Putt's favorite cleaning adage in mind: Prevention is the best cure. My kitchen sink almost never gets clogged and that's because every morning after I make my cup of tea, I pour the rest of the boiling water right down the drain. The hot water dissolves any oil or grease buildup in the drain, which prevents most clogs. Also, I keep a drain basket in my sink at all times. Of course, that keeps anything but water from going down my drain. When I do get a clog, I turn to my favorite method: I pour two-thirds of a cup of baking soda into the drain. I follow that with the same amount of white vinegar, and then I cover the drain with a plate. The bubbling action of vinegar and baking soda dissolves many clogs.

Homemade Dishwasher Detergent

> 1 part baking soda
>
> 1 part borax

If there's one thing that we can all agree on, it's that an automatic dishwasher makes life a heck of a lot easier than washing every last dish by hand. But if you think you're paying too much for store-bought dishwasher detergent, replace it with this inexpensive mixture you can use for every load.

DIRECTIONS: Combine the two ingredients and funnel the powder into a clean, dry container with a lid (like a recycled plastic gallon milk jug). If you have hard water, double the amount of baking soda in the mixture. Use 2 tablespoons of the detergent for each load of dishes. And for a spot-free shine, pour in ½ cup of white vinegar at the start of the rinse cycle.

Easy Does It 🏠 HERE'S A QUICK AND EASY way to make greasy grime and musty odors inside your dishwasher disappear like magic. Just pour 1 cup of white vinegar into a sturdy mug or bowl, set it upright in the top rack, and run through a cycle on the hottest setting—with nothing else inside.

Q *My dishwasher has a lingering, unpleasant odor, even right after it finishes a cycle. I've tried wiping out the interior with cleansers, but the smell lingers. What can I do to get rid of it?*

A Start by sprinkling 3 tablespoons of baking soda in the bottom of the empty machine, let it sit overnight to neutralize the odor, and wash it away by running the dishwasher through a regular cycle. If that doesn't do the trick, wipe the area around the drain with damp paper towels sprinkled with a little baking soda. This will remove any greasy residue that may be decaying there. Do the same around the rubber gasket that is either around the door or around the face of the unit where the door meets it. If you can still smell a problem, then you'll need to find the source. Get down on your hands and knees, and look closely at the inside of the dishwasher. Wipe out any grime using a gentle abrasive, such as baking soda. If your dishwasher still stinks, call in a plumber—the problem could be a clogged pipe.

Inside/Outside Fridge Fixer

> 1 part water
> 1 part white vinegar

Use this simple but powerful solution to clean your fridge from stem to stern, including the gasket around the door. It'll remove dirt in a jiffy, kill nasty germs, wipe out lingering odors, and make mold move on out!

DIRECTIONS: Combine the ingredients in a handheld sprayer bottle, spritz everything in sight, and wipe the grime away with a sponge or soft cloth. There's no need to rinse; the smell of the vinegar will disappear as it dries. And don't forget to do the top of the refrigerator, too, even if you can't see the dust up there!

Inside Wipe 🏠 AFTER YOU REMOVE the shelves and bins, wipe down the inside of your refrigerator with a clean, damp cloth that's been sprinkled with a little baking soda. The soda provides a gentle abrasive for scrubbing off any stuck-on bits of food, plus it deodorizes while it cleans. Rinse off the residue with another clean, damp cloth and you'll be ready to reload.

Brighten Up 🏠 TO ELIMINATE fruit juice, cola, ketchup, and other stains from inside a fridge, squeeze a dab of white non-gel toothpaste on the spots and rub it in with a damp cloth. Rinse well with another damp cloth when you're done, and your refrigerator will be smiling brightly!

PINCHING PENNIES

Be extra careful when you clean glass or plastic refrigerator shelves because replacements cost a pretty penny! To clean the shelves thoroughly, remove them and wipe both sides with my Inside/Outside Fridge Fixer (above). If they're loaded with hardened crud, let them soak for a few minutes in lukewarm, soapy water before you rub the residue off. And never wash a cold glass shelf with hot water: The sudden change in temperature could crack the glass. Once they're cleaned and rinsed, let the shelves air-dry before you put them back into place.

Kitchen Cupboard Cleanup Formula

Sure, "shabby chic" cupboards can make your kitchen feel as comfortable as an old shoe. But if your cozy cabinets are looking a little too "lived in," perk 'em up with this deodorizing solution.

DIRECTIONS: Combine the ingredients in a bucket, then use a sturdy sponge to wash your cupboards—inside and out. Rinse your sponge with clear water between swipes, so the stuff in your bucket stays potent.

Cut the Grease 🔺 IF YOUR KITCHEN cabinets are feeling a little sticky, then you've got greasy buildup. The oily residue from cooking can quickly dull even the brightest cabinet's shine, so you need to wipe down those doors at least once a month. The simplest solution? Just use a damp sponge with a dab of grease-cutting dishwashing liquid on it. Rinse with a clean, damp sponge, and dry with a lint-free microfiber cloth to make the finish look as good as new.

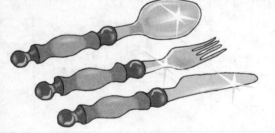

Shine Silverware 🔺 WANT TO MAKE your silverware (and glassware) sparkle like new? Add a capful of bleach to a sink full of water and your regular dishwashing liquid, and that'll do the trick!

Super Shortcuts

✂ To coax stubborn cabinet crud into giving up its grip, grab a bottle of Spray 'n Wash® stain remover from your laundry room, and give the surfaces a spritz. Wait about a minute or so to let the foam do its stuff, then finish up by wiping off the stain remover and the greasy dirt with a damp sponge. If your cabinets are made of wood, play it safe and test this technique in an inconspicuous place first to make sure it doesn't affect that beautiful finish.

Kitchen Grease Buster

When you're cooking for a crowd (or if you're just a really messy eater), spills and splatters can build up on every surface you touch. For those extra-greasy, tough-to-clean areas around the stove, mix up this cleaning potion.

> ¼ cup of white vinegar
> 2 cups of hot water
> 1 tsp. of borax

DIRECTIONS: Pour the vinegar and water into a handheld sprayer bottle, then spoon in the borax, put the lid on the bottle, and give it a swirl to mix the ingredients. Spritz the mixture on the tough-to-tackle trouble spots, and wipe it away with a damp cloth.

Tough Customers  TO REMOVE hardened globs of oatmeal, pancake batter, or other stubborn foods from your stove top, first saturate the spots with water, and let them sit for about 10 minutes. Then use a nonabrasive plastic scrubbie to loosen the stuck-on food, and wipe the residue off with a damp sponge. Whatever you do, don't use steel wool pads or scouring powders on stove tops because they can easily scratch the surface.

It Pays to Get Fresh STOVE TOP SPILLS are easier to clean up when they're fresh, so wipe up spatters as soon as you see them. If a spill has already dried, lay a wet paper towel over it after the stove is cool, and wait about 10 minutes for the water to soften it up. Then simply wipe it clean.

That's Brilliant!

To save yourself a whole lot of messy cooking cleanup work, use heavy-duty aluminum foil. Simply remove the drip pans and line each one with foil, trimming it to fit with a pair of scissors. Then the next time a pot boils over, all you have to do is replace the foil liner. For even more protection, use foil to line the drip tray under your cooktop, too.

Lunch Box Stink Sinker

A kid (or, let's face it, some adults) can definitely put a lunch box through the wringer, carrying it back and forth every day. With all the wear and tear, something is bound to spill inside, and odors can permeate the plastic (or whatever material it's made of). Perk up a stale-smelling lunch box with this vinegar "sandwich."

DIRECTIONS: Soak the bread in the vinegar, then place it inside the lunch box. Seal the lunch box and leave it alone overnight. In the morning, remove the soggy bread, and wipe the inside of the lunch box clean with a wet sponge.

First-Aid Lunch 🏠 A LUNCH BOX is just the right size for a first-aid kit. So if you've got a lunch box that's out of work, stock it with supplies like bandages, antibiotic ointment, and pain relievers, and stash it where you can grab it fast when you need it. (Of course, if it's a vintage collectible decorated with pictures of, say, Davy Crockett or Zorro, keep it out in the open to be "Wowed!" at.)

Odor Absorber 🏠 TO SOAK UP nasty smells inside of a soft-sided lunch box, sprinkle some baking soda in it, making sure you get it into the folds in the lining, where odors may be hiding. Zip it closed and let it sit overnight. After the baking soda has had a chance to absorb the odors, wash and dry the container, and it'll be ready to be put to good use. By the way, this works well in good old-fashioned metal lunch boxes, too. Just pour baking soda inside the box and shut it tight for the night.

PINCHING PENNIES

To get rid of bacteria that may be contributing to the smell inside of a lunch box, spray it with a solution made of 1 tablespoon of bleach and 1 quart of water. The chlorine will kill germs instantly. Wash the solution away with a warm, soapy cloth after you spray it on, and the box will be food-safe for next time.

Microwave Gunk Buster

> ½ cup of baking soda
> 2 tsp. of white vinegar
> 4 drops of lemon oil

If you don't wipe your microwave clean after each and every use (whether you think it needs it or not), you're going to end up with sticky, grimy globs on the inside walls, and some pretty nasty odors, to boot. So bust the gunk with this everyday cleaner, and keep in mind that you should never, ever use abrasives or metal scrapers to clean the inside of your microwave!

DIRECTIONS: Combine all of the ingredients in a bowl, adding more vinegar or baking soda as needed to make a thick paste. Slather the walls of the microwave with the mixture and let it sit for a minute or two. Swipe the paste away with a damp sponge, and follow up with a thorough warm-water rinse.

Spray and Zap 🏠 HERE'S A QUICK and easy way to clean a dirty microwave oven. Spray the inside lightly with plain water, being careful not to direct the spray into the vents, close the door, and zap it on high for 60 seconds. Then open the door, and wipe away the crud. An alternative method is to wet a dishcloth and zap it on high for 60 seconds. Let it cool off for three or four minutes with the door closed, so the steam loosens the grunge. Then wipe down the inside of the oven with the same cloth. It doesn't get any easier than this!

Q *The last time I cooked fish in my microwave, the oven still stank three days later! Is there anything I can do the next time to get rid of that fishy smell quickly?*

A There sure is. When smelly foods leave their aroma behind after you microwave them, use lemon water to get rid of the stink. Fill a microwave-safe bowl halfway with water, drop in a wedge or two of fresh-cut lemon, and zap it on high for two minutes. Keep the door closed for five minutes while the water steams and cools down a bit. Remove the bowl, dip a cloth into the lemon water, and wipe down the interior. This lemony rinse will leave a nice fresh scent behind.

No-Rinse Surface Scrub

Kids keep a home lively, that's for sure. And they'll happily play with whatever they have on hand—even food. When little ones are around, sticky fingers, gunky food splashes, and smeary grime seem to wind up on your kitchen's surfaces no matter how careful you are. So grin and bear it, and swipe them away with this surefire solution.

> 1 cup of baking soda
> 1 cup of lemon-scented ammonia
> 1 gal. of warm water

DIRECTIONS: Combine the ingredients thoroughly in a large bucket. Then, wearing rubber gloves, apply the mixture to smeared and spattered walls, cupboards, and drawer fronts with a sponge. Scrub gently to quickly erase any marks. That's all there is to it—you won't even need to rinse!

Stone Solutions 🏠 IT'S EASY TO LOVE the look of countertop materials like soapstone, sandstone, and Jerusalem stone, but the ongoing care could be more than you bargained for. These materials can scratch easily, so make sure you always use a cutting board or silicone sheet to do your chopping and slicing. And because grease stains can be absorbed by these natural stones (just as they can by granite or slate), keep a cloth handy whenever you're working with oily foods. That way, you can promptly wipe up greasy spills and residue before they have a chance to sink in.

Super Shortcuts

The cleaning power of fresh-cut lemons is legendary—the citric acid they contain is the bleaching agent that makes lemons such handy helpers around the house. But if you don't have time to cut up lemons for everyday cleaning chores, bottled lemon juice works just as well. Mix lemon juice with baking soda to clean coffee or tea stains on laminate countertops. And when mixed with salt, lemon juice can remove even a tough stain like a red wine spill on a tablecloth.

Oak-Kay Cabinet Cleaner

Take special care when cleaning your oak kitchen cabinets. If you use too much water or saturate the surfaces with a liquid cleaner, your woodwork is likely to end up looking worse than it did when it was simply dirty. This gentle cleaner will have your oak looking okeydokey, and leave your whole kitchen smelling clean.

DIRECTIONS: Combine the Murphy Oil Soap with the water in a bucket, then add the patchouli essential oil. Carefully pour the mixture into a handheld sprayer bottle, and spray sparingly on your cabinets. Use a soft cloth to wipe the woodwork clean. Your cabinets will shine, shine, shine!

*Available at health-food stores.

Welcome to the Club 🏠 FIZZY

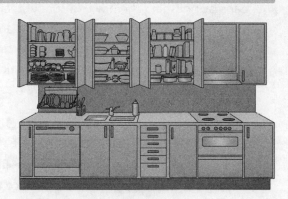

CLUB SODA breaks right through that dull, greasy buildup on cabinet surfaces. So the next time you have a little extra left in the bottle, don't let it go flat in the fridge! Instead, use it to dampen a cleaning cloth and wipe the grime away.

That's Brilliant!

Have your cabinets been neglected for far too long? If so, then here's the best way to whip them back into shape. First, remove all the handles and knobs from the cabinets and drawers. Set the screws aside, and put the hardware in a sink filled with hot water and a few squirts of grease-cutting dishwashing liquid. Let the soap soak away the grime while you clean the cabinet surfaces—a task that's much easier without having to work around the handles. Then wash each handle with a sponge, and reattach it. Your once-grungy cabinets will have a new lease on life.

Old-Time Aluminum Cleaner

Grandma Putt kept her aluminum pots and pans looking brand-spanking new with this simple homemade cleanser, and so will you. (It'll also work its magic on your aluminum outdoor furniture!)

½ cup of baking soda
½ cup of cream of tartar
½ cup of white vinegar
¼ cup of soap flakes
    (such as Ivory Snow®)

DIRECTIONS: Combine the baking soda and cream of tartar in a bowl. Add the vinegar and mix to form a paste. Stir in the soap flakes, transfer the mixture to a glass jar with a tight-fitting lid, and label it. Apply the paste with a plain steel wool pad, and rinse with clear water.

Restore the Shine 🏠 GET RID OF the cloudy look of your aluminum pots and pans by polishing them with a soapy steel wool scouring pad. To give the surfaces a uniform shine, rub them using a back-and-forth motion instead of going around

and around in circles. Or you can sprinkle some baking soda onto the wet surface and rub—back and forth!—with a nonabrasive plastic scrubbie to polish that dulling film away.

Boil Away Stains 🏠 CLEAN STUBBORN COOKING STAINS from the inside of an aluminum pan by placing it on the stove, filling it about halfway with water, and adding 2 to 3 tablespoons of lemon juice to it. Bring the liquid to a rolling boil, then reduce the heat to medium-high and let the acid in the lemon juice simmer away those tough stains for about 15 minutes or so.

Super Shortcuts

✂ A stain can linger even after you scrub the burned-on food out of an aluminum pan, but you can bring your cookware back to brightness with this easy, time-saving trick. Just fill the pan with water, drop in two Alka-Seltzer® tablets, and let it soak for an hour. What a relief it is to be rid of those lingering stains from burned-on sauces or spaghetti!

Perfect Pewter Polish

Pewter is mostly tin, so it needs a lot of TLC because it can be easily scratched or damaged. Never use steel wool or harsh abrasive cleansers, and don't even think about setting it on a hot stove or putting it in the dishwasher. For a safe cleaning solution, try this mildly abrasive paste. It's easy on pewter, but tough enough to tackle the tarnish.

DIRECTIONS: Mix the vinegar and salt, then add enough flour to make a smooth paste. Smear the mixture onto the pewter piece, and allow it to dry for 30 minutes or so. Rinse the pewter with warm water, and polish the item with a soft cloth. You may need to use a cotton swab to get the paste out of any grooves. **Note:** As you polish the pewter, make sure you rub it back and forth in one direction and not in circles, so your strokes blend in with the metal.

Avoid a Melting Moment ⬟ UNLIKE HARDIER METALS, pewter begins to melt at only 450°F, so never heat your pewter pieces or set them in the oven or microwave. And pewter must always be washed by hand—the hot water in your dishwasher will overheat the precious metal.

Please Handle the Merchandise ⬟ THE MORE YOU touch your pewter, the quicker it will pick up a patina, which is what gives the finish a beautiful antique appearance. The oxidation will also help protect the piece over time.

GRANDMA'S OLD-TIME TIPS

Old-timers like my Grandma Putt had a terrific trick for polishing their pewter—they used cabbage leaves! Just peel off a few leaves from a head of regular cabbage and rub down the surface of your pewter pieces. The leaves will remove the grime and restore a gentle gleam without scratching the metal.

Range Hood De-Greaser

1 cup of white vinegar
½ cup of baking soda
2 cups of hot water

Oily molecules that go airborne when you're cooking seem to crash land on the range hood, hardening into a layer that takes tons of elbow grease to remove. To keep the crud to a minimum, wipe your hood down every week or so with this simple spray. It'll cut the greasy residue, and save you from a bigger cleanup job later.

DIRECTIONS: Mix the ingredients in a handheld sprayer bottle, spritz the potion on the range hood (inside and out), and wipe away the greasy film. Rinse off the residue with a damp cloth, and your hood will be good to go. You can also use this degreaser on your stove top and other large appliances.

Ammonia Power
WHEN A RANGE hood is extremely dirty, call on the grease-cutting power of ammonia to eat through the oily buildup. Simply add a squirt of dishwashing liquid and ¼ cup of ammonia to a bucket of hot water, dip in a sponge, and rub the grime away. Rinse with a damp cloth, and you'll really be cooking!

When Not to Clean Your Range Hood Filter
IF YOUR RANGE HOOD is not vented to the outside of your house, it's equipped with a charcoal filter. While you can—and should—wash the filter cover in hot, soapy water periodically, the filter itself cannot be cleaned. You will need to buy a new one every six to nine months, depending on how often you use your stove top. (Check your owner's manual for specific guidelines.)

Super Shortcuts

If your range hood filter has been acting up or accumulating gunky buildup, it most likely needs a thorough cleaning. But don't spend time scrubbing away at it. Instead, lay the filter screen in the bottom of a heavy-duty trash bag and pour 1 cup of ammonia over it. Tie the bag tightly, and let it sit outside (so it doesn't smell up your house) for 24 hours. Once the ammonia's worked its magic, rinse the filter off with a garden hose, and put it back in place.

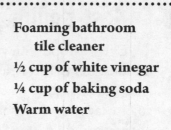

Refrigerator Bubble Bath

When it's time to clean out the fridge, don't be afraid to get tough. Start by clearing everything off the shelves and door, and dump any food that's expired or any open containers filled with stuff you know you're not going to consume. Remove the shelves and wash them by hand in the sink or according to the manufacturer's directions. Then tackle the refrigerator's interior using this cleanser.

Foaming bathroom
tile cleaner
½ cup of white vinegar
¼ cup of baking soda
Warm water

DIRECTIONS: Spray the interior walls of the fridge with the foaming bathroom tile cleaner. Then measure the vinegar and baking soda into a handheld sprayer bottle, and fill the remainder of the bottle with warm water. Spray this mixture over the foaming bathroom cleaner and wipe the walls clean with a large sponge, making sure you rinse the sponge frequently. Dry the interior with an old absorbent bath towel, and you'll be ready to return your cold goods to a spotlessly clean appliance.

Stuck-On Spills 🏠 TO LOOSEN UP food residue inside a fridge, heat ½ cup of white or cider vinegar in a microwave or on the stove top, pour it into a small, heat-proof bowl, and set it in the fridge for about five minutes. The steaming vinegar will unstick those stuck-on spills, making it easy to wipe them away. The inside of your fridge will look like new again, and any smells that were accumulating will be long gone.

PINCHING PENNIES

Dirty condenser coils will make your fridge work harder to keep cool, and that costs money. So you need to vacuum underneath the fridge and around the back every now and again to remove any loose dirt. Then every few months, do a more thorough job by removing the cover panel at the bottom and vacuuming out any debris.

Rust-Routin' Recipe

Ugly rust spots on your cookware can make it look older than it is, and can even get bits of rust into your food when you cook. You can soak away rusty spots on steel, iron, or enamel cookware with very little elbow grease when you rustle up this terrific tonic.

> 1 part white vinegar or
> ¼ part lemon juice
> 4 parts water
> Vegetable oil

DIRECTIONS: Fill a pan with the vinegar or lemon juice and the water so that the rusty spot is covered. Let the pieces soak overnight, and if there's any rust remaining the next morning, scrub it off with a nonabrasive plastic scrubbie or a steel wool pad. Rub the spots with a light coating of vegetable oil to head off further problems, and you'll really be cooking.

Plastic for Stainless SCRATCHES ON STAINLESS STEEL cookware have a nasty habit of causing food to stick, which makes cleanup a lot harder the next time. So avoid rough scouring powders and steel wool pads that may leave their mark on the surface, and when it's time to get rid of cooked-on crud, reach for a nonabrasive plastic scrubbie instead. You'll be glad that you did!

Foiled Again WHEN YOUR OUTDOOR grill or patio furniture gets rusty, scrub the corrosion away with a ball of crumpled aluminum foil. Then keep rust at bay by rubbing on a coat of clear car wax so rain can't penetrate the finish.

That's Brilliant!

If your stainless steel pots and pans seem truly hopeless, here's a neat tip that'll bring them back to life, fast. This little trick will not only save your cookware, but it'll keep you from having to waste a lot of energy scrubbing and scrubbing (with little result). Remove any non–stainless steel handles or knobs, and put the pots and pans in your oven while you run the self-cleaning cycle. The cooked-on grime will burn to ash, just like it does in your oven!

Silver Soak 'n' Shine

Restore the gleam to your tarnished or dull silver or silver-plated flatware by soaking it in this solution. It works like crazy, thanks to a chemical reaction that dissolves the tarnish even in tiny crevices, leaving each piece clean and bright. **Note:** Don't use this potion if you want to preserve the patina in the flowers, scrolls, or other intricate designs on your flatware!

> 1 gal. of warm water
> 4 tbsp. of table salt
> 4 tbsp. of washing soda*
> Aluminum foil

DIRECTIONS: Mix the water, salt, and washing soda in a large glass pan, and stir until the salt and washing soda are dissolved. Lay a sheet of aluminum foil in the bottom of the pan, and drop in your silver-plated flatware. Let it soak for about an hour, then rinse thoroughly, and dry each piece with a clean, soft cloth.

*Available online and in some hardware stores.

Sink or Dishwasher? SILVER AND SILVER-PLATED flatware actually benefits from hand washing and drying because the gentle rubbing helps deepen the patina. Stainless steel, on the other hand, stays its shiniest when you use the dishwasher to clean it. But if you decide to hand wash your stainless and it's looking a little dull or water-spotted, moisten a cloth with white vinegar and wipe it clean.

Follow the Lines WHENEVER POLISHING SILVER, use a straight back-and-forth motion and not a circular one. That way, the strokes will blend into the surface. And use a light touch when you rub, letting the polish—not your elbow grease—do the job.

GRANDMA'S OLD-TIME TIPS

Grandma Putt knew when certain types of flatware absolutely had to be washed by hand. One of those times is when your silver or silver-plated flatware has hollow handles. Don't let these stay submerged for more than a few minutes because soapy water can loosen the solder on the pieces. Remember—handle your hollowware with care!

Smudge-Free Stainless Solution

If you're like most folks, you love the look of stainless appliances in your kitchen, but you have a devil of a time keeping them smudge-free. Here's an easy way to wipe away fingerprints, grime, and gunk, and leave your stainless as it should be—stainless!

DIRECTIONS: Don't mix the ingredients together. Instead, put a few drops of baby oil on a lint-free cloth, and wipe the entire surface of the stainless appliance. Then moisten a clean cloth with club soda and use it to rinse off the oil, and wipe everything dry with another clean cloth.

Soap's the Solution  TO BUST RIGHT THROUGH stubborn crud on stainless steel, dip a cloth into a solution of grease-cutting dishwashing liquid and water. Dishwashing liquid is gentle on stainless but tough on dirt, so it does a terrific job on greasy stove tops and grimy refrigerators. Be sure to rinse the surface thoroughly after you wipe away the grime, because soap residue can leave a rainbow hue on stainless steel. Then use a lint-free towel to give your stainless a nice shine.

Cut the Grease IF IT'S BEEN A WHILE since you wiped down your stainless steel kitchen appliances, use one of the homemade stainless steel scrubs in this chapter instead of a commercial cleaner. They work just as well at cutting through greasy film, and cost a whole lot less. Regardless of which potion you use, be sure to rinse the cleaner off with fresh water, and dry the surface to a sparkling shine.

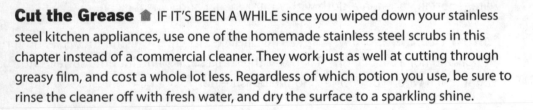

That's Brilliant!

Scratching the surface of stainless appliances, sinks, or cookware is a big no-no because it damages the protective layer that keeps the steel from rusting. To keep your finishes in fine fettle, avoid using steel wool and other abrasive pads, as well as gritty scouring powder. So how do you get stuck-on food off the bottom of stainless pots and pans? A good soak and a nonabrasive plastic scrubbie are all it takes.

Stain-Chasin' Stainless Scrub

Here's a quick homemade cleaner that's safe for stainless steel sinks, cookware, and other surfaces. It does a great job of removing stains and discolorations, so the beauty of the stainless can shine through.

> 3 parts cream of tartar
> 1 part hydrogen peroxide
> Rubbing alcohol

DIRECTIONS: Mix the cream of tartar and hydrogen peroxide in a glass bowl, and apply the cleaner with a soft, damp cloth. Let it dry, and then wipe the residue off with a clean, damp cloth. Next, wipe the surface with the rubbing alcohol, and then buff it dry with a clean, soft cloth. Use long, even strokes, working with the grain every step of the way, and you'll be pleased with the sparkling-clean results.

Oil Slick ♦ TO GIVE CLEAN STAINLESS STEEL a super shine, rub the surface with a bit of oil. Olive oil or food-grade mineral oil works best because neither of these will turn rancid or create a film, as other cooking oils might do. Rub the oil over the entire surface and remove any excess with a soft, dry cloth. The very thin layer of oil that's left behind will add a gleam and protect the surface from water spots and stains.

Hair's an Idea ♦ WANT TO MAKE YOUR STAINLESS steel sink sparkle like diamonds? Then mosey into your bathroom, grab a bottle of hair conditioner, and pour a dab of it onto a clean, soft cloth. Wipe the sink with it, then stand back and admire the glow.

Q My stainless steel sink got blue streaks after I set a hot saucepan in it—and so did the bottom of the pan. What can I do to get rid of the blues?

A Those blue marks are heat stains because, as you've discovered, stainless steel discolors when it gets extra hot. To make heat stains disappear, rub them with club soda or olive oil on a soft cloth until they're gone. And next time, don't set hot cookware in the sink until it cools down, and never let the liquid in a stainless pan boil away.

Super Silver Shine Solution

The shinier silver pieces are, the more beautiful they appear. But keeping silver constantly at its brightest is harder than you'd think. Here's the easiest recipe I know of for getting tarnished silver bright and shiny again.

> 2 pans that can hold enough water to cover your silver piece
> Aluminum foil
> 1 cup of baking soda per gallon of water
> Water

DIRECTIONS: Line the bottom of one pan with foil, and set in your tarnished treasure. Make sure the silver touches the aluminum (I say this in case there's a non-silver part to the object). Then fill the second pan with water and heat it to boiling. Remove the pan from the heat, set it in the sink, and add the baking soda. Be careful—the solution will foam up and may spill over. Pour the soda solution into the first pan, completely covering the silver. Within seconds, you'll see the tarnish start to disappear. A lightly tarnished piece should be clean as a whistle in four or five minutes; one with a heavy coat of tarnish may need a few more treatments.

Acid Indigestion 🏠 FOODS THAT CONTAIN ACID can discolor silver when they react with the metal, and salty foods can bring on the tarnish, too. That's why many silver serving pieces come with a glass liner. If your serving pieces aren't protected by a liner, don't use them to serve acidic foods like fruit juice, eggs, mayonnaise, and vinegar salad dressing, or salty items like olives. You can also search online for "glass liners for silver." You may get lucky and find some that will fit your pieces.

GRANDMA'S OLD-TIME TIPS

Whenever my Grandma Putt boiled a pot of potatoes, she saved the water—not for soup, but to clean her silver. She simply soaked her silver earrings and other small pieces in the potato water for about 30 minutes and then dipped a soft cloth in the water to rub the tarnish away. It worked like a charm!

Tea-rific Teapot Tonic

There's nothing like a spot of tea to warm you up on a chilly morning. If your favorite Darjeeling has left its mark on your teapot, well, that's not anyone's cup of tea. But you'll be back to stain-free sippin' after removing the evidence with this formula.

> 1½ cups of apple cider vinegar
> 3 tbsp. of salt
> 1½ cups of water

DIRECTIONS: Combine the ingredients in your teapot, then bring the mixture to a boil. Let it boil for 15 minutes or so, then let the pot continue to soak for 24 hours. Rinse the pot out, and it'll be all set to brew.

Teapot Cleaner 🏠 NO MATTER HOW WELL you wash a teapot after each use, those brown tannin stains still build up over time. But you can remove them by filling the pot with boiling water, tossing in a handful of borax (roughly ½ cup), and letting it sit overnight. Then give the pot a good washing before making your next cuppa.

Ceramic Cleaner 🏠 GET ANYTHING MADE OF CERAMIC squeaky clean by wiping it with ¼ cup of white vinegar mixed in a bucket of warm water. From heirloom teapots to tile floors, it'll shine all your ceramic surfaces with a few good swipes.

Q *I have an old teapot that belonged to my grandmother, so it has sentimental value. Problem is, the antique china pot has a hairline crack. I can't use it to brew tea, but can you think of another way I can put it to good use?*

FAQ?

A I sure can. When I was a young lad, I remember seeing one of my Grandma Putt's favorite heirloom teapots on display. It too had a few minor cracks, so Grandma turned it into a twine holder. She just popped the ball inside the pot and threaded the end through the spout. Then anytime she needed to tie up a package or truss a turkey, she pulled out as much twine as she needed and snipped it off.

Teflon®-Tough Pot Cleaner

Your Teflon pots are pretty darn tough, but even their resilient coated surface can get baked on and caked on too many times and lose its luster. Use this solution to ditch the dirt and keep the nonstick surface ready for action.

> 1 cup of white vinegar
> ¼ cup of baking soda
> 2 cups of water

DIRECTIONS: Place all of the ingredients into the Teflon pot and boil the mixture over high heat for 10 to 15 minutes. Remove the pot from the stove, dump the contents into the sink, and give the pot a quick wash with soapy water. Rinse the pot under hot water and allow it to dry completely. Season the Teflon surface by swiping it with a light coating of vegetable oil on a paper towel.

Nonstick Fix TO LOOSEN STUCK-ON CRUD from nonstick cookware, pour in 1 part baking soda to 1 part water, using just enough to cover the bottom of the pan. When the solution turns brown, lift off the food residue with a plastic spatula, and wash the pan as usual. It's the simplest way to keep those pans true to their word and stick-free.

GRANDMA'S OLD-TIME TIPS

While Grandma never had any cookware that was this fancy, she still had a remedy that works great on it. To loosen baked-on food from CorningWare®, fill the dish with a solution of 1 part white vinegar to 3 parts water, and let it soak for several hours. Or you can cover the baked-on residue with a paste of baking soda and a little warm water, and let it sit overnight. Remove the residue with a nonabrasive plastic scrubbie. And to remove any gray "scratches" on your cookware, use scouring powder—the scratches are actually metal left behind when your utensils scrape against the hard CorningWare surface.

Ultimate Range Remedy

3 tbsp. of baking soda

3 tbsp. of salt

Pinch of cream of tartar

Hydrogen peroxide

It seems like the range gets dirty faster than it takes to whip up a one-dish dinner. As soon as you get one spill cleaned up, two more appear. It's a never-ending cycle of grime and buildup. When your stove top is covered in pancake batter, gravy splatters, and who knows what else, whip up a batch of this cleaner.

DIRECTIONS: Combine the baking soda, salt, and cream of tartar in a bowl. Add enough hydrogen peroxide to make a paste. Spread this concoction on the stains and let it sit for a half hour or so, then scrub with a sponge dipped in warm water.

Filthy Fan ⌂ AFTER YOU REMOVE the range hood filter for cleaning, take a few minutes to spiff up the fan that's inside the unit. Unplug it if you can, and then dust the blades with a microfiber cleaning cloth. Once the loose dirt is gone, spray some cleaning solution on a damp cloth and wipe the blades clean. Check out the fan compartment; if it's really dusty, use the crevice tool of your vacuum cleaner to suck the dirt away. That way, your range hood will be fresh and clean, both inside and out.

Soak the Small Stuff ⌂ TO CLEAN YOUR STOVE'S knobs and drip pans, simply pull them off and put them in a pan filled with a gallon of warm water and ¼ cup of dishwasher detergent (*not* dishwashing liquid). Let them soak while you clean the rest of the stove (or longer if they're really dirty). Use a toothbrush to scrub off any stubborn spots. Then rinse everything with clear water.

Super Shortcuts

The dishwasher can make short work of cleaning the metal range hood filter that covers the exhaust fan. Remove the filter, put it in the top rack of the dishwasher, and run it through the pots and pans cycle. The extra-hot water will scrub it clean, and you won't have to lift a finger. If the plastic or glass covering of the range hood's light is attached to the filter, leave it in place and it, too, will come out sparkling clean.

What's Cookin' Oven Cleaner

Sure you can clean your oven with chemicals, but let's face it, your oven—not to mention your entire kitchen—will smell like chemicals for a long time! Try this all-natural cleaner instead. It'll do a tip-top job and leave behind a nose-pleasin' smell in its place.

> 2 cups of baking soda
> ½ cup of salt
> ¼ cup of borax
> Water
> 1 cup of white vinegar
> 5 drops of lemon oil
> 5 drops of thyme oil

DIRECTIONS: Remove the racks from your oven, and set the temperature to 225°F. While the oven heats up, mix together the baking soda, salt, and borax in a bowl. Add enough water to make a thick paste. After the oven has heated for 15 minutes, turn it off and open the door. Carefully spread the paste on the walls of the oven and leave it alone for at least half an hour. Then mix the vinegar and essential oils in a handheld sprayer bottle, and spray the mixture evenly over the oven walls, saturating the paste. Leave it alone for about 15 minutes, then wipe the oven clean, grime and all, using a sturdy cloth and rinsing frequently. Ahh! It's like a breath of fresh air.

Heavy-Duty Oven Cleaner IF YOUR OVEN hasn't been cleaned in a long time, don't despair. Give this potent potion a try. In a small bucket, mix 1 teaspoon of dishwashing liquid, 1 teaspoon of lemon juice, and 1½ teaspoons of bleach into 1 quart of warm water. Use a sponge to wipe down the interior of your oven with the solution, and let it sit for 45 minutes before scrubbing clean with a nonabrasive plastic scrubbie. Then rinse well with warm water.

That's Brilliant!

Don't throw away your leftover tea leaves! Instead, save them to make this tea-rific oven cleanser. Combine 1 cup of black tea leaves, 1 teaspoon of baking soda, and a few drops of dishwashing liquid. Heat the mixture, and wipe it all over your oven's interior. Let it sit for a few minutes, scrub with a brush or sponge, and rinse with warm water.

Wooden Cutting Board Deodorizer

Baking soda
Salt
Water

A wooden cutting board is a great addition to any kitchen. But sometimes the board needs an extra boost to clear out the smells of the onions, peppers, and garlic you chopped. To keep a wooden cutting board clean and deodorized, no matter what's cookin', try this terrific tonic.

DIRECTIONS: Mix equal parts of baking soda and salt with enough water to make a paste. Scrub this concoction into the board, leave it on for a few hours, and then rinse thoroughly. This will remove the odor and leave your cutting board ready to use next time.

When It Rains, It Pours! 🏠 WHEN THE SURFACE of your wooden cutting board gets mottled by fruit juice or other stains, make those marks disappear by sprinkling salt on the stains and rubbing the board with a freshly cut lemon wedge. If you need stronger abrasive action than ordinary table salt, try using coarse kosher salt instead. And keep in mind—the sooner you wipe up a spill, the fewer stains you'll have to deal with.

How Dry I Am 🏠 GERMS AND MOLD can set in fast on damp surfaces, so make sure you dry cutting boards thoroughly after washing them. Use paper towels to dry the surface of the boards, then lean the cutting boards upright to air-dry before you put them away.

GRANDMA'S OLD-TIME TIPS

Grandma Putt had a little trick for keeping her wooden cutting boards clean as a whistle. She would rub some food-grade mineral oil into them every so often. If you try this, let the oil soak in for about 10 minutes, and then wipe off the excess with a paper towel. The wood will absorb the oil, and that'll help keep germs from making themselves at home in the board.

CHAPTER 2

BATHROOM
Brighteners

Some folks think they need an arsenal of commercial cleaners to really get their bathroom sanitized. If you're one of them, have I got news for you—you can stop spending your cash on fancy bottles of store-bought concoctions. Instead, reach for the baking soda, vinegar, and a few other common household products. I'll show you how to use these simple, family-friendly formulas to get your bathroom fresh and sparkling clean without blowing your budget.

Amazing Ammonia Cleaner

¼ cup of ammonia
2 gal. of water

Plain old ammonia is one of the best cleaners around because it can clean just about any surface. This mix gets rid of grime like magic. But remember, folks—ammonia is powerful stuff with powerful fumes, to boot! So always handle it with care and crack open a window before you begin.

DIRECTIONS: Mix the ingredients together in a sturdy plastic 3-gallon pail, then just dip a sponge or a cloth into it to clean windows, mirrors, and anything else that you'd use its commercial counterpart, Windex®, for. (Because this homemade version is way cheaper than the store-bought stuff, you can even use it outside the bathroom for jobs big and small, from spiffing up appliances to cleaning dingy aluminum siding.) Just remember to wear a pair of rubber gloves whenever you work with this potion. And never, ever, *ever* mix it with bleach—the resulting fumes can be deadly!

Vinegar for a Shiny Sink

HARD-WATER BUILDUP in your bathroom sink can be the very dickens to remove, but not if you use a little white vinegar. Just close the drain, pour in some vinegar, and let it sit overnight. Come morning, open the drain and wipe the sink to a shine.

Refresh Your Fixtures

MAKE YOUR CHROME FIXTURES sparkle by lightly spraying them with seltzer and wiping with a soft, dry cloth. There's no elbow grease required—just stand back and admire the shine!

Super Shortcuts

Shampoo will do a bang-up job of cleaning grime from your shower walls. The same product that's so good at cutting through the dirt and oils in your hair will do the same for built-up scum on the tiles and glass door, leaving them sparkling clean and shiny. So lather up your hair, rinse, and repeat the whole process on your shower's surfaces!

Bathroom Shine-Up Solution

1 part white vinegar
1 part water

Here's one of my favorite Earth-friendly cleaners: a simple half-and-half mixture of vinegar and water. Use it to remove dirt and add shine to countertops, windows, mirrors, and other hard, glossy surfaces. **Note:** If your bathroom has marble countertops, fixtures, or decorative objects, don't use this cleaner! The acid in vinegar—even in a drift of spray—can permanently damage your precious stone.

DIRECTIONS: Combine the ingredients in a handheld sprayer bottle and shake the mixture before using. Then spray it on and wipe the grunge away; there's no need to rinse because the sharp vinegar scent will dissipate as it dries. You can keep this potion on hand indefinitely in a cool, dry cabinet. If you use this solution to clean mirrors or windows, add a few drops of dishwashing liquid to the mix to prevent streaks.

From Fridge to Faucet BELIEVE IT OR NOT, a plain old ordinary bottle of ketchup will work wonders to clean brass or bronze faucets. Just squirt a big dab onto a dry cloth, and rub it over the faucet until all of the grime is gone. Then rinse the faucet and buff it dry. And don't blame me if you suddenly get a hankering for a nice juicy burger and a pile of fries . . . with plenty of ketchup, of course!

Know Your Onion Bags HERE'S A DIY SCRUBBER that'll really help you clean up: Clip off the end of an empty onion bag and stuff it with other empty onion bags. Then wad it up and use it to tackle soap scum on shower walls, and even grimy grout.

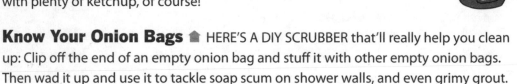

That's Brilliant!

It's tough enough to keep mirrors clean, but when you throw hair spray into the mix, it's darn near impossible. To get rid of that built-up hair spray haze that tends to attach itself to bathroom mirrors, just dab on a little rubbing alcohol, then wipe with a soft cloth. Now you're looking *really* good!

Better Than Baking Soda Bathroom Cleanser

½ cup of ammonia
¼ cup of baking soda
Warm water

A quick sprinkle of baking soda on a damp sponge is usually enough to swipe away simple bathroom soap scum. But why not add some power to the powder, and give your cleaner a kick that'll disinfect as it cleans? This cleanser will deep clean your bathroom and leave it sparkling like the summer sun.

DIRECTIONS: Mix the ammonia and baking soda in a handheld sprayer bottle, then add enough warm water to fill the bottle. Keep this formula on hand for everyday bathroom cleanup. Be sure to wear rubber gloves, and use it in a well-ventilated area. And if you just can't handle the smell of ammonia, replace it with ¼ cup of dishwashing liquid, and increase the amount of baking soda to 1 cup.

Candle Wax Cleanup 🛁 THERE'S NOTHING LIKE taking a warm, soothing bath in candlelight, but when candle wax has splashed or dripped on your bathroom wall, it sure can ruin the mood. Don't let it: Whether your wall is painted or papered, just heat it with a hair dryer turned on the highest setting, and blot the melted wax up with a paper towel. Keep heating and blotting until no more wax or oil soaks into the wall. If the candle wax left a stain behind, use a Mr. Clean® Magic Eraser® on the spot until it's gone. This trick works great for waxy crayon marks, too!

PINCHING PENNIES

The best way to keep marble countertops looking good is to avoid using cleaning products on them altogether. Plain water is all you really need. But if there's greasy residue on the surface, add a few drops of mild dishwashing liquid to a bucket of warm water. Wash off the oily film and rinse with clean water right away. Then dry the marble with a soft chamois, so the water doesn't spot the stone.

Budget-Pleasing Denture Soak

You say your pearly whites aren't looking so pearly? Baking soda will work wonders to restore dazzle to your dentures. So mix up a batch of this magical solution for your mouth!

> 2 tsp. of
>   baking soda
> 1 cup of
>   warm water

DIRECTIONS: Stir the baking soda into the warm water until it dissolves, add your dentures (or your kids' retainers!), and let them soak until you need them. The baking soda will loosen food particles and keep coffee and cola stains from settling in. Plus, it'll neutralize any food odors, so your dentures will taste as fresh as they look. Simply rinse off the solution when you're ready to wear your nice fresh teeth again.

Chopper-Calming Tea 🏠 HERE'S A QUICK (and tasty) way to ease denture pain. Drop a teaspoon of dried chamomile into a cup of hot water and steep for 10 to 20 minutes. Strain out the herb, and when the tea is cool, sip a mouthful, swish it around for 30 seconds or so, and spit it out. Keep rinsing until you've used all of the tea.

Denture Ache 🏠 IF YOUR NEW DENTURES or braces are a little bit uncomfortable, toughen up your gums by gargling with a solution of 1 to 2 teaspoons of salt in a glass of warm water. (But if your mouth is still tender after a few days of this treatment, call your dentist or orthodontist.)

GRANDMA'S OLD-TIME TIPS

Grandma always knew the best way to clean everything, including dentures. Every few days, she'd wash her dentures with a soft, soapy cloth, making sure to cover the entire surface, both inside and out. To try this technique, place a rubber mat in the sink before you get started, so your teeth will make a soft landing if they happen to slip out of your hands. And be sure to rinse your choppers thoroughly when you're done, so they aren't flavored with soap when you slip them back in.

Classic Comb and Brush Cleaner

¼ cup of ammonia
¼ cup of dishwashing liquid
2 cups of water

Take a good look at your comb and brush. Is that a "Yuck!" I hear? All that filth comes from hair products, natural oils secreted from your scalp, and strands of loose hair. So spiff 'em up with this terrific tonic.

DIRECTIONS: Mix the ingredients in a large bowl, then let the combs and brushes soak for 5 to 10 minutes. Remove them from the solution, and scrape the brushes with the combs, then vice versa. Rinse them with cool water.

Got Oily Hair? REMOVE STUBBORN GRIME from your comb by giving it a grease-cutting bath. Just dunk it in a glass filled with white vinegar to cut through the greasy residue, or squirt some shampoo for oily hair onto your comb, and soak it in a glass of warm water for an hour. Then rinse and wipe it dry with a paper towel.

Clean Your Teeth COMBS CAN GET mighty gunky, especially if you use sprays or gels on your hair, which then end up as a sticky film on your comb. So keep those teeth sparkling clean by soaking your comb in hot, sudsy water about once a week. Begin by pulling out any loose hair, and then let your comb soak for a couple of hours to soften the buildup. You'll then be able to wipe away most of the grunge with a paper towel. For any lingering crud, use an old toothbrush to scrub the last of it out of the comb's teeth.

Super Shortcuts

Why clean only your comb when you can save time, money, and effort by using your comb and brush to clean each other? First, pull out any and all loose hairs and lint. Next, wet your brush and comb, squirt some shampoo directly onto the brush, and scrub them together, face-to-face. The teeth of the comb will clean the brush, and vice versa. Rinse out all traces of soap under warm running water, and dab out any remaining dirt with a cotton swab. And if you have two hairbrushes, scrub them together, brush-to-brush—to clean both at the same time!

Clear the Air Germ Killer

1 cup of water
½ cup of dried eucalyptus leaves
½ cup of rubbing alcohol
2 tsp. of eucalyptus oil

Let's face it: There are times when a bathroom smells just awful. Here's an herbal disinfectant spray that will perk it up with a minty-fresh aroma and knock out any lingering germs.

DIRECTIONS: In a small saucepan, bring the water to a boil and add the eucalyptus leaves. Reduce the heat and simmer for 15 minutes. Let the mixture cool, then strain out the herbs and pour the liquid infusion into a glass spray bottle. Add the alcohol and essential oil. Tighten the spray cap and shake well. Spray into the air as needed. This elixir will last up to six months if stored in a cool, dry place.

The Sweet Smell of Clean 🔼 ADD A PLEASANT scent to your scrubbin', shinin', and sanitizin' by mixing 1 cup of baking soda with 20 drops of your favorite essential oil. Let the mixture sit for about 24 hours, then sprinkle a little of the scented powder mix into the sink, the bathtub, the toilet, or wherever you want to clean. It makes surfaces spic-and-span and smells great, too!

PINCHING PENNIES

If you can't seem to rid your bathroom of mildew—and the telltale aroma that goes along with it—it's time to wage war against moisture and humidity. Start by filling one foot of an old pair of panty hose with a scoop of unscented clay cat litter, knot the hose at the ankle, and cut off the extra part above the knot. Then place this nylon sack on a shelf in the bathroom to absorb odor and moisture from the air. And remember to always turn the ventilation fan on before you step into the shower or take a steamy bath, and let it run until the bathroom air loses its extra humidity.

Come Clean Comb Bath

> 2/3 cup of white vinegar
> 2 drops of tea tree oil
> 2 cups of warm water

Gels, sprays, shines, and mousses—when it comes to hair care, there's more than enough stuff to gunk up a comb even without the usual lint, loose hair, and general grime. So make an appointment to have your teeth cleaned—your comb's teeth, that is!

DIRECTIONS: Mix the ingredients in a bowl, and completely immerse your comb. Let it sit in the solution for at least 30 minutes, then rinse the comb under warm water and place it on an absorbent towel to air-dry.

Clean for Clean 🏠 SCRUB YOUR BRUSH AND COMB before you shampoo your hair, not after you style it. That way, you'll be using clean tools on your clean hair—and that greasy brush won't dull your locks.

Baking Soda Bath 🏠 BUBBLE AWAY COMB DIRT with good old baking soda. Here's how: Fill your sink with a couple of inches of hot water, and pour in about a cup of baking soda. Submerge your combs, letting them soak for half an hour. Then wipe off the grime, and rinse.

Give Old Hairbrushes the Brush-Off 🏠 WHEN YOUR HAIRBRUSHES get too ragged to use on your hair, get 'em out of your bathroom—but don't throw them away. Instead, move them to the laundry room or broom closet. Then put them to good use cleaning the inside of your bagless vacuum cleaner, your dryer's lint trap, and any other hard-to-get-at places.

That's Brilliant!

Plastic hairbrushes and combs can soak up the perfumes from styling products and the odor of oily hair and general dirt. So freshen them up by soaking them in a sink full of warm water with a squirt of shampoo and 1/2 cup of borax added to it. Swish your brushes and combs in the mixture, then let them soak for an hour. Finish by rinsing them well.

Company's Coming Grout Scrub

> 2 cups of baking soda
> 1 cup of borax
> 1 cup of hot water

When guests come over, having a clean bathroom can make all the difference between your home looking lovely and inviting and your home looking like a junkyard with a big old "not welcome" sign hanging on your door. If your bathroom grout isn't looking so great, use this potion to spiff it up in a hurry—it's safe to use on both the white and colored stuff.

DIRECTIONS: Mix the ingredients in a small bucket. Dip a nonabrasive plastic scrubbie or grout brush into the mixture, and rub it briskly along the lines. Rinse thoroughly with a cloth or mop dipped in plain water to remove the residue.

Trash to Treasure 🏠 THE NEXT TIME YOU CLEAN your bathroom, reach for a plastic mesh produce bag (the kind that onions and potatoes come in). It makes a perfect bathroom scrubber, whether you're using rubbing alcohol or any other cleanser. It won't scratch porcelain, tile, or marble surfaces, yet it's tough enough to clear even the most stubborn grime—like mildew and soap scum—off grout.

Scrub Out Grout 🏠 TO CLEAN THE GROUT between ceramic tiles, wet it down with a cloth or sponge, then dip a damp toothbrush in baking soda, and scrub the dirt away.

PINCHING PENNIES

Are you spending an arm and a leg on bathroom deodorizers? Well, Grandma Putt would say, "Stop that right now!" Instead, when an, um, aromatic incident leaves the room smelling less than rosy, just light a candle. Don't bother searching for a scented one. Any candle flame will burn away the foul smell.

Daisy-Fresh Spruce-Up Formula

½ cup of white vinegar
1 tsp. of borax
1 tsp. of liquid castile soap
1 tsp. of washing soda*
3 ½ cups of hot water

With daily splashes, splatters, and soap scum, your bathroom starts looking grungy pretty darn fast. This hardworking formula gets walls and countertops sparkling clean lickety-split—and it leaves behind a fresh, clean scent, too!

DIRECTIONS: Mix the ingredients in a handheld sprayer bottle, then shake well. Spray non-wood surfaces and wipe them clean with a damp sponge. Your entire bathroom will be gleaming in no time at all.

*Available online and at some hardware stores.

Toilet Ring 🏠 HERE'S HOW TO MAKE a stubborn toilet bowl ring vanish: Combine just enough borax and lemon juice to form a paste, and cover the ring. Let it sit for an hour, then scrub with a toilet brush.

Ring Glider 🏠 SOME MORNINGS YOU PULL and pull on your shower curtain, but it just won't slide. You have to wiggle and jiggle it until it finally closes. Here's how to make those curtain rings slide smoothly and end your morning frustrations: Smear a thin coat of petroleum jelly onto the rod. That's all there is to it!

Keep Yer Flap Happy 🏠 WHEN HARD-WATER MINERALS build up in a toilet tank, the deposits keep the flapper valve from closing tight—and that leads to an annoying trickle. If jiggling the handle doesn't make the toilet stop running, shut off the water, flush the toilet to empty the tank, and scrub off the buildup with a heavy-duty scouring powder. Then turn the water back on and flush the toilet to rinse out the residue.

Super Shortcuts

✂ If you get knee pain while scrubbing the bathroom floor, put on a pair of knee pads (found in the volleyball section of a sporting-goods store) before you get down to business. Your knees will say "thank you!"

Do-It-Yourself Liquid Hand Soap

2 cups of mild laundry soap flakes

10 oz. of hot water

2 tbsp. of baby oil or mineral oil

It's oh-so-handy to keep liquid soap by the sink—and the hand pump makes it a whole lot neater than keeping a slippery, sudsy bar there. But it sure can get pricey, especially if you have a large family. With this recipe you can save yourself a buck or two, and refill the same pump bottle over and over again.

DIRECTIONS: Put the soap flakes in a bowl and stir in the water. Add the oil and make sure the entire concoction is well blended. Pour the mixture into an old liquid soap dispenser or cheap pump bottle. The soap will tend to separate, so give it a good shake every once in a while before you use it. If you prefer scented soap, you can add a few drops of your favorite fragrant essential oil.

Homemade Towelettes 🏠 MOIST TOWELETTES CAN BE lifesavers when you're away from home and you make a mess. But there's no need to waste money on store-bought wipes—just make your own. Mix a few tablespoons of baby oil with a cup of soapy water. Soak some paper towels (the kind that won't fall apart when wet) in the solution, and store them in individual plastic snack bags. Stash a couple in your purse and in your car's glove compartment for quick cleanup whenever you need it.

PINCHING PENNIES

As you well know, cleaning the bathroom (and housework in general) can make your hands really rough. But the cost of those store-bought, super-duper hand lotions can put your budget in bad shape! Here's a make-it-yourself concoction that'll leave your hands soft and smooth, while keeping your wallet in your pocket. Combine 2 tablespoons of dried chamomile flowers, 1 tablespoon of dried rosemary, and 4 cups of water in a sauce pot. Boil, uncovered, over medium heat for about 10 minutes. Cool to room temperature, and then strain. Pour the liquid into a bottle and refrigerate. Apply with a cotton ball (or, if you have a spray bottle, you can spray it on instead), and let your hands air-dry.

Dynamo Drain De-Gunker

Lots of gunk winds up going down bathroom drains. From hair and whiskers to toothpaste and shaving cream—not to mention soap suds and other residue—all that crud starts backing up the drain in no time flat. So flush out your bathroom sink and shower drains once a month with this mighty mix.

1 part baking soda
1 part salt
¼ part cream of tartar
½ gal. of boiling water

DIRECTIONS: Mix the dry ingredients and pour ½ cup into the drain. Dump the boiling water on top of the dry mix. Then run hot water from the faucet for several minutes, followed by cold water for several more minutes. This will keep the toothpaste, hair, and other nasties from clogging up your drains.

Odor Remover 🏠 DIRTY DRAINS LEAD TO bad smells, so you'll want to keep yours squeaky clean and smelling fresh. And the easiest way to do that is to pour 1 cup of vinegar or bottled lemon juice in them once a week. Then let it sit for 30 minutes, so the mild acid can cut through the gunk. Finish up by running hot water for a minute or so to flush the drains. No more stinky sinks!

That's Brilliant!

You can clear a clogged drain easily with Alka-Seltzer® Original antacid tablets. Just drop two tablets into standing water and let the effervescent fizz do all the work. After about 30 minutes, flush the drain with water. Your drain will be on the go—er, flow—just like that! And if you don't have any antacid tablets handy, hydrogen peroxide can cut the clog, too. Pour ¼ cup of 3% hydrogen peroxide down the drain. If there's standing water in the sink, carefully use your plunger to clear the gunk away. Otherwise, flush the drain with water and it'll be flowing freely in no time at all.

Easy All-Surface Cleaning Elixir

2 cups of rubbing alcohol
1 tbsp. of ammonia
1 tbsp. of dishwashing liquid
2 qts. of water

Manufacturers and product pitchmen would have you believe that you need a different (expensive) cleaner for every type of surface in your home. Well, that may be true to a certain degree, but here's a great recipe for a multipurpose cleaner that can be used in your bathroom and just about anywhere else.

DIRECTIONS: Combine all of the ingredients in a handheld sprayer bottle, and go to town. (Incidentally, this super-duper concoction will beat commercial, streakless glass-cleaning products hands down!)

Bust the Rust ⬢ WHEN YOU FIRST BRING metal cans of shaving cream and hair spray home, coat the bottoms with clear nail polish. That way, you'll no longer have to clean up those ugly rust rings these types of cans tend to leave on countertops and bathtubs. Don't have any clear nail polish on hand? No problem! Wrap the bottom of the can with a piece of plastic wrap and hold it in place with a rubber band. The beauty of this solution is that you can use the wrap over and over.

FAQ?

Q *I spilled nail polish on my bathroom's white marble countertop. I've scrubbed and scrubbed, but the stain is still there. Any advice?*

A When dye from a nail polish spill seeps into a white marble surface, your best bet is to spread a cleaning paste over the stain to draw out the dye. Mix 1 tablespoon of hydrogen peroxide and 1 tablespoon of water with enough powdered laundry detergent to make a paste, and smear it onto the stain. Cover the poultice with a damp cloth, and let it sit until the stain has been bleached away. Then wipe the paste off with a moist cloth, and polish the marble with a soft, dry cloth.

Everyday Scum-Away Shower Spray

½ cup of hydrogen peroxide

½ cup of rubbing alcohol

2 tsp. of spot-free liquid dishwasher detergent

2 to 3 drops of grease-cutting dishwashing liquid

3 cups of water

Preventing dirt from building up in the first place sure beats scrubbing it off later! So use this spray after your daily shower to ward off the scum on the walls and in the tub. It'll make your next full-scale cleaning a heck of a lot easier.

DIRECTIONS: Mix all of the ingredients in a handheld sprayer bottle with a "mist" setting on the nozzle, and shake to combine. Then park this miracle mixture right on the tub, where it will be handy after your daily shower. Once you're fresh and clean, take a minute to mist the walls of the shower, so that blasted soap scum never gets a chance to settle in. Hard-water spots will be foiled, too, because this solution dries fast and streak-free.

So Long, Soap Scum! THIS QUICK FIX will make that cloudy layer disappear from the shower wall in a jiffy. Just sprinkle a little baking soda on a damp sponge, rub away the cloudy film, and then rinse thoroughly to remove all traces of the residue.

Showerhead Cleaner CLEAR THE OPENINGS of a clogged showerhead by filling a large ziplock plastic bag with white vinegar and attaching it to the shower-head with rubber bands. After an hour or so, remove the bag and test the flow. If some holes remain blocked, open them up with a toothbrush, toothpick, or darning needle.

That's Brilliant!

Has your showerhead gone from the bathroom version of Niagara Falls to a feeble trickle? Hard-water deposits are probably clogging the holes. To keep your shower flowing freely, fill your sink with hot water and add a couple of denture-cleaning tablets. Then unscrew the showerhead and soak it in the mixture for several hours. Rinse it well, screw it back in place, and you'll enjoy a full-force shower again.

Fix It Fast Drain De-Clogger

Take it from me—even bald folks get clogged bathroom sinks from time to time. Shaving cream, soap scum, and toothpaste can make your drains S-L-O-W. When hair is not the culprit, de-clog your drain and keep it running freely by chasing away any residue with this powdery ploy.

> ½ cup of baking soda
> ½ cup of table salt
> 1 ½ tablespoons of cream of tartar
> 2 cups of cold water

DIRECTIONS: Combine the baking soda, salt, and cream of tartar, then pour the mixture down the drain. Follow up with the cold water. Repeat the steps if necessary, until water flows freely down the drain.

Denture Drain Cleaner 🏠 SPEED UP A SLOW DRAIN by using denture-cleaning tablets to clean the pipes. Just drop three tablets down the drain, pour in a cup of white vinegar, and wait about 10 minutes. Run hot water to flush the drain, and watch that water flow!

Smelly Drain Solution 🏠 TO FRESHEN UP a smelly drain, pour a cup of baking soda into it, and flush with hot water. The baking soda will neutralize odors in the drain and leave your bathroom smelling a whole lot better in no time at all.

Grandma's Old-Time Tips

When your drains are so clogged up that water never seems to drain completely, try my Grandma Putt's quick fix, and plunge the problem away. Grab a "plumber's friend"—a hand plunger—hold it over the drain, and then give it one or two quick, hard pumps to dislodge the gunk that's causing the backup. Wear long rubber gloves for this job because you'll need to quickly grab the hair ball or other goop before it slides back down the drain. Then keep your drains clean by regularly using my Fix It Fast Drain De-Clogger recipe (above).

Fresh as a Summer Day Disinfectant

Clean your bathroom and clear the air. And while you're at it, give your nose something to smile about. This herbal sanitizing spray is sure to liven things up!

DIRECTIONS: Place all of the essential oils and water in a medium bowl and mix thoroughly. Then pour the mixture into a glass jar with a lid, and seal it tightly. When you need to freshen things up in the bathroom, pour a small amount into a handheld sprayer bottle, and go for it.

> 1 tsp. of lavender oil
> 1 tsp. of lemon oil
> 1 tsp. of rosemary oil
> ½ tsp. of rose oil
> ¼ tsp. of peppermint oil
> ⅛ tsp. of clove bud oil
> 1 cup of distilled water

All-Purpose Disinfectant

HERE'S A MULTIPURPOSE disinfectant recipe that you can use on everything from bathroom countertops and toilets to baby bottoms! Just combine the following ingredients in a handheld sprayer bottle: 2 cups of water, 2 tablespoons of dishwashing liquid, and 30 drops of tea tree oil. Shake the bottle to mix the contents well, then take aim at those dirty surfaces.

Mouthwash Germ Killer

ANTISEPTIC MOUTHWASH KILLS GERMS anywhere. If you need to clean the toilet bowl and you're fresh out of other cleansers, just reach for the mouthwash, pour some onto a sponge, and clean away!

FRESH BREATH

Super Shortcuts

The built-up ring that shows the high-water mark in your tub and sink can sure be tough to get rid of, even with a lot of added elbow grease. If scrubbing doesn't seem to work, try soaking. Soak white rags or paper towels in household bleach and lay them on the ring. Let them sit for about an hour, then rinse to your heart's content. The rags keep the bleach in place on the stain—allowing it to do its job instead of simply running down the drain.

Great Grout De-Gritter

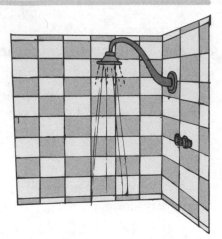

> 3 cups of rubbing alcohol
> 2 cups of bleach
> ½ cup of liquid floor cleaner
> 1 qt. of water

Sometimes when you wipe off a dirty tiled wall, it just doesn't seem as clean as it should. That's probably because the extra-stubborn dirt and grime got stuck in the grout and refused to budge. But even the grimiest grout will come clean as a whistle with this homemade "miracle" spray.

DIRECTIONS: Mix the ingredients together in a bucket. Then pour the solution into a handheld sprayer bottle, and use it as you would any spray cleaner. Store the leftover mixture in a tightly sealed container, well out of reach of children and pets.

Gel in the Shower 🏠 BECAUSE IT'S SO THICK, gel-formula toilet bowl cleaner with bleach does a great job of cleaning shower wall grout, even in hard-to-reach areas. Start at the top of the tile wall and squeeze the gel on the grout lines, letting it slowly run down to the shower floor. The cleaner will do its dirty work in a few minutes, and then you can simply rinse it off. Now, wasn't that easy?

PINCHING PENNIES

As they say, "When in Rome, do as the Romans do." I say, when in the bathroom, use something from the bathroom to clean that ugly, stained grout. That's my roundabout way of telling you to save those old toothbrushes—they're the perfect-sized tools for this scrubbing job. When it's time to clean stained grout, you could also use a little whitening toothpaste to get the job done lickety-split, but that's a pretty pricey solution. Instead, whip up some homemade whitening paste by combining hydrogen peroxide with baking soda to make your own "scrubbing bubbles." This concoction works pretty well on your teeth, too; just remember to use a different toothbrush!

Homemade Baby Wipes

Baby wipes aren't just for babies. Many folks like to keep 'em on hand in the bathroom and use them to freshen up after—ahem—using the facilites. If you like babying your bottom, don't spend your money on the store-bought variety when you've probably got everything you need to make your own. Here's how to make a whole stack.

- 1 roll of soft, absorbent paper towels (premium brands work best)
- 1 plastic container with a tight-fitting lid to hold the paper towels
- 2 tbsp. of baby oil
- 2 tbsp. of liquid baby bath soap
- 2 cups of water

DIRECTIONS: Cut the roll of paper towels in half with a serrated knife, and remove the cardboard tube. Place half the roll, on end, in the plastic container. Mix the liquid ingredients, pour the solution into the container, and close the lid. The towels will absorb the liquid. As you need them, pull the wipes up from the center of the roll.

Wonder Wipe THERE ARE A ZILLION AND ONE uses for baby wipes! Here are just a few bathroom-related tips: Use 'em to polish bathroom surfaces, clean cuts and scrapes, and remove stubborn makeup. They also come in handy for a quick cleanup when you aren't near soap and water.

That's Brilliant!

Plastic baby wipe boxes are tailor-made for storing all kinds of odds and ends in the bathroom. They're perfect for corralling cotton balls, cotton swabs, bobby pins, hair clips, ponytail holders, or what have you. If you collect soap slivers to combine later (as well you should!), these boxes are great containers to hold them in. They also come in handy for first-aid supplies. (I keep kits in my bathroom, kitchen, work-shop, car, and garden shed, just in case.)

Magical Mirror Mix

> 1 cup of rubbing alcohol
> 1 cup of water
> 1 tbsp. of white vinegar

Your teenager just spent an hour in front of the mirror, and now you have all kinds of streaks and grime smudging it up. Your mirror will shine with this simple mixture that tackles fingerprints, makeup smears, hair spray, greasy residue, and other gunk with just a few quick swipes.

DIRECTIONS: Pour the ingredients into a handheld sprayer bottle, and shake well before using this mix. Spray the potion onto a microfiber cloth, and wipe the dirt away. There's no need to rinse, and the solution will dry quickly, thanks to the alcohol in it. **Note:** This solution keeps indefinitely in a cool, dark cabinet.

Check Behind the Mirror 🏠 NEXT TIME YOU'RE CLEANING your bathroom mirrors, take a minute to open up the medicine cabinet and give its contents the once-over. If you're hanging on to half-empty bottles of pills the doc prescribed in 1999, it's time to dump 'em! Medications go downhill fast, so chances are they've completely lost their potency by now. And even if there is some oomph left in those meds, they could harm you if you take them. Toss your old cosmetics while you're at it, too. Even though makeup companies are not required by law to show expiration dates on their labels, cosmetics do deteriorate over time. So if you still have that tube of "Pretty in Pink" that you wore to your senior prom, do yourself a favor and pitch it!

GRANDMA'S OLD-TIME TIPS

Whenever Grandma noticed scratches in the bathroom mirror, she didn't even think about replacing it or sending it out to be repaired. She fixed it right at home, and it only cost a few pennies. Here's how: Rub a little dab of white non-gel toothpaste into the mark, let the paste dry, then buff the area with a soft, clean cloth. The excess paste will wipe away, but the scratch will remain filled.

Move Out Mildew Spray

½ cup of rubbing alcohol
1 tbsp. of liquid laundry
    detergent with enzymes
3 cups of water

One of the biggest bad guys in the shower is mildew. It's stubborn, and it always seems to keep popping up when you least expect it. If you're tired of battling the mildew that builds up on your bathtub, glass shower doors, or vinyl shower curtain liner, then here's a simple formula that'll stop the slime before it has a chance to form.

DIRECTIONS: Mix all of the ingredients in a handheld sprayer bottle, and keep it on the side of the tub. Then issue an all-points bulletin that says the last person out of the shower or bath each day must spray the solution on all of the wet surfaces. Follow up once a month by wiping down the walls with the same solution, and you can kiss mold and mildew good-bye.

Polish the Walls 🏠 EVEN THOUGH OILY RESIDUE can be a real mess in the bathtub, it's one solution to soap scum on shower walls. A very light layer of oil will actually slow down the formation of stubborn scum. Try a liquid lemon furniture polish for a pleasantly scented preventive—just pour a small amount on a soft, dry cloth, and lightly rub it over the shower walls in a circular motion. Or if citrus isn't your thing, use baby oil instead. Just make sure you don't get any oil on the floor, or you'll have a slippery situation the next time you step into the shower.

Super Shortcuts

If you want to save yourself a lot of cleaning time and elbow grease, invest in a chamois. A super-soft and absorbent chamois (either the leather or synthetic version) makes a perfect cleaning aid because it soaks up water in two shakes of a lamb's tail. Just wipe down the wet walls of the shower and tub to remove soap residue and hard-water spots. Then you won't have to worry about cleaning that shower nearly as often.

Odor-Chasing Citrus Spray

This super solution will freshen the bathroom without all the chemicals you'd find in commercial air fresheners. Plus, the light, fresh scent will clear the air instead of masking one bad odor with an even stronger one.

> 1 cup of distilled water
> 1 cup of rubbing alcohol
> 2 drops of lemon oil

DIRECTIONS: Combine the ingredients in a handheld sprayer bottle and spritz the mixture a few times wherever things tend to get a little stinky.

Bathroom Freshener ⬆ IF YOUR BATHROOM still has a lingering, stale smell from your morning shower, there's a simple way to make it springtime fresh. Just apply a drop or two of lemon extract to the lightbulbs in the bathroom fixtures, *before* you turn them on. (Feel free to use peppermint, almond, or vanilla extract instead.)

All-Purpose Bathroom Cleaner ⬆ MAKE AN ALL-PURPOSE "miracle" cleaner by mixing 1 part lemon juice with 2 parts water in a handheld sprayer bottle. Spray the solution on surfaces in the bathroom, and wipe grime away with a nylon-covered sponge or a mesh onion bag.

Warm Up the Radiator ⬆ TO MAKE A RADIATOR or baseboard heater work more efficiently in your bathroom, wrap a piece of heavy-duty aluminum foil, shiny side up, around a piece of cardboard or wood, and tuck it behind the unit. The foil will reflect heat into the room, so it won't be absorbed into the wall. And you won't catch a chill when you step out of the shower or tub.

Q *My bathroom windows are looking downright dirty, and they're making my bathroom smell awful. Is there anything I can use to clean and deodorize them that's not a commercial product?*

A If your bathroom windows are really grimy, then wash them with a solution of 1/2 cup of white vinegar and 2 tablespoons of lemon juice in 1 quart of warm water. It works like a charm—and smells like a daisy—every time!

Over-Easy Orange Sanitizing Solution

> **3 cups of white vinegar**
> **1 orange peel**

There's a reason citrus-scented cleaners sell so well. Their pleasant aroma makes not only your bathroom, but also the whole house just a little nicer. So if you love the smell of orange disinfecting cleaners, but aren't so keen on the price, then make a bottle of the stuff yourself!

DIRECTIONS: Combine the ingredients in a jar with a tightly closing lid, then set it aside for two weeks. Remove the peel and strain the vinegar, then pour it into a handheld sprayer bottle to clean the surfaces in your bathroom (and kitchen). Your rooms will smell as good as they look. You can also clean linoleum floors with this cleaner. Just add 1 cup of it to 2 gallons of water, and mop away.

Zesty! BUILT-UP SOAP SCUM and hard-water deposits are no match for the grime-fighting power of lemons. The citric acid works like crazy to clean up tiles, tubs, and shower stalls, and the lemon-fresh smell just can't be beat. Juice six lemons, strain out the pulp, and pour the remaining liquid into a handheld sprayer bottle. Spray it onto the scummy surface and give the lemons a few minutes to do the dirty work. Then rinse well with warm water so there's no sticky residue, and wipe the surface dry with a soft absorbent cloth or towel.

That's Brilliant!

Oranges aren't just good for making bathroom cleaners. The oil in orange peel is 90 to 95 percent limonene, a natural chemical that's been linked to preventing breast and cervical cancers, at least in test tubes. Limonene ends up in the OJ you buy because commercial machines squeeze the oranges so hard. So, while you're making your own cleaner, have a glass of OJ and get your dose of limonene.

Porcelain Sink Perk-Up Potion

½ cup of white vinegar
¼ cup of baking soda
10 drops of rose oil

Porcelain sinks are beautiful—unless they're marked up with stains, scum, and splotches. If you have one in your bathroom, you know how great it can look. So use this gentle cleaner to super-clean your porcelain. It'll tackle the tough stuff with no surface-scratching abrasives.

DIRECTIONS: Mix the vinegar, baking soda, and essential oil together in a bowl, then dip a damp sponge into the solution and use it to clean the sink. Rinse with clear water, and repeat if necessary. If you still see grime, add ¼ cup of borax to the mixture, and clean the sink again.

Sparkling Sink GET A CERAMIC SINK SPARKLING clean by filling it with 2 to 3 inches of water, then dropping in two Alka-Seltzer® tablets. Wait 20 minutes or so, then drain the sink and wipe it down with a damp sponge. It'll sparkle like new.

Porcelain Sink Whitener WHITEN UP THAT PORCELAIN sink by filling it with a solution of ¾ cup of bleach per gallon of water. Let it sit for five minutes, and rinse thoroughly. Your porcelain will be cleaner and brighter than it's been in years!

PINCHING PENNIES

Say you decided to be a little more "manly" with your bathroom and went with the stainless steel sink instead of one made of porcelain or ceramic. Well, just because it's easier to clean doesn't mean that it doesn't need its own special treatment, just like the others. You can make a stainless steel sink sparkle by using something that should already be in your bathroom—hair conditioner. Just use a dab of conditioner on a soft cloth, and rub it into the steel until it shines.

Safe-and-Sound Bathroom Drain Cleaner

> 1 cup of baking soda
> 1 cup of salt
> ½ cup of white vinegar
> 2 qts. of boiling water

Soap scum, toothpaste gunk, shaving cream, and hair can make a yucky mess in tub and sink drains. But you don't need toxic chemicals to clean the stuff out. This gentle formula will have those pipes flowing freely again in no time at all.

DIRECTIONS: Combine the baking soda, salt, and vinegar, and pour the mixture down the drain. Wait 15 minutes, then follow with the boiling water. Turn on the hot water, and let it run into the drain for one minute. For really stubborn clogs, repeat until the water is flowing freely.

Don't Mix and Match ⬆ BAKING SODA AND OTHER household products can react explosively with commercial drain cleaners, so never, *ever* use them before or after using a commercial product. And always play it safe by not combining any treatments, whether they're commercial or homemade.

The Dynamic Duo ⬆ YOU CAN CUT A LOT OF CLUTTER from your bathroom—and make it a much safer place, to boot—if you replace your collection of commercial cleansers and cleaners with two basic kitchen staples: a bottle of vinegar and a box of baking soda. Stash them wherever you keep your bathroom cleaning supplies, whether that's under the sink, or in your linen closet or utility room. Performing together, or in solo roles, these two superstars can keep your bathroom spic-and-span.

Q *With three long-haired girls in our family, it seems like every other day I'm struggling with a clogged sink or bathtub drain. How can I keep my bathroom drains from getting clogged all the time?*

A Keep your drains free and clear by using a drain strainer to trap hair so you don't find yourself digging it out months from now. In the meantime, use a plunger to break up a clog and allow it to float away.

Sharp-Shooter Shower Spray

Shower tiles and grout seem to stay wet—no matter how quickly and thoroughly you dry the walls after each use. This fast-acting formula makes it as easy as pie to keep your tub, shower, and tile in tip-top shape.

½ cup of white vinegar
1 tbsp. of borax
1 tbsp. of dishwashing liquid
3 cups of hot water

DIRECTIONS: Mix all of the ingredients in a handheld sprayer bottle, and give it a good shake. When you're ready to tackle that shower, just spray it down, and then wipe the cleaner (and the crud) away.

Make Your Shower Sparkle 🏠 CLEANING THE SHOWER tends to be the most tedious job in the bathroom. But don't worry—if you follow my technique, it'll be a breeze. Start by spraying the shower walls and glass doors with tile cleaner, like my Sharp-Shooter Shower Spray (above). Scrub the tiles and the grout between them with a tile brush. Next, scrub the doors with a nonabrasive plastic scrubbie. Use a toothbrush and the spray cleaner to get the grime off the tracks of the shower doors. Then scrub the bottom and sides of the tub, using a tile brush and the right product for your type of tub. Don't forget to clean the joint where the tub meets the shower wall with a toothbrush. Then rinse the shower walls and doors thoroughly, followed by the tub. Finally, clean the outside of the shower doors and outside of the tub with an all-purpose spray cleaner. That's all there is to it!

PINCHING PENNIES

Remove stubborn water spots on your chrome faucet by rubbing it with a fresh-cut lemon half, letting the juice soak in for a few minutes before you rinse it with fresh water. Then go ahead, take a great big sniff—now that's the smell of clean! And to make that shine last longer, wipe the faucet with a dab of baby oil on a soft cloth. That way, the water will bead up on the thin film of oil instead of on the metal, and any spots that do show up will be easy to wipe away.

Shower Curtain Liner Bath

Laundry detergent
½ cup of baking soda
1 cup of white vinegar

When everything in your bathroom has to be cleaned—from the toilet to the tub, and everything in between—don't even think about sponging down your shower curtain liner to get it clean. It's much easier to remove it from the curtain rod and toss it in your washing machine. And with this easy tonic, the liner will look neat and nifty, just like that!

DIRECTIONS: Place the liner in your washing machine, along with two or three bath towels. Add the normal amount of laundry detergent for your machine and the baking soda. Set the water temperature to warm and let 'er rip. Add the vinegar to the cycle during the final rinse. Go ahead and toss the towels in the dryer, but take your liner back to the bathroom and hang it on the curtain rod to air-dry.

Conquer the Cling

ELIMINATE ANNOYING STATIC CLING on a vinyl shower curtain liner by mixing 1 capful of fabric softener in 2 cups of water in a handheld sprayer bottle, and spritzing the liner from top to bottom. If the curtain itself is vinyl, spray it, too. Now your curtain and liner will open and close smoothly as you step in and out of the shower.

Super Shortcuts

If your bathroom mini blinds are a grungy mess, reach for a can of foaming bathtub cleaner and spray away, so those scrubbing bubbles can lift off the crud. Wipe the surface with a damp sponge, then pull the cord to reverse the slats, and do it again. Just be careful where you spray that stuff—keep it on the blinds and away from any painted or wood surfaces. Oh, and if those pull cords are dirty, too (and they probably are), give them a shot of shaving cream and a rubdown with a damp sponge to freshen them up.

Super-Sanitizing Wipe

Use this potion on any hard surface that needs a quick cleaning. One swift swipe will cut through grime and kill germs at the same time!

> 1 part rubbing alcohol
> 4 parts water

DIRECTIONS: Mix the ingredients in a handheld sprayer bottle. Then simply spray the solution on the surface to be cleaned, and wipe it away with a soft, clean cloth. Keep a bottle of this solution on hand in the bathroom for quick counter and mirror cleanups, and keep another bottle in the kitchen to swipe appliances, faucets, and sinks clean.

Fingerprint Kit 🏠 TO REMOVE SMUDGES and dirty fingerprints from painted or papered walls in your bathroom, rub the spots gently with an art gum eraser. You can also use a dry rubber sponge, which is available at paint-supply stores. It'll erase the grunge from your walls lickety-split.

White or Rye? 🏠 IF THE WALLPAPER in your bathroom isn't washable, but still gets pretty darn dirty, you don't have to live with it. Instead, you can use a couple slices of fresh bread to rub the dirt away. Simply wad the bread into a fist-size ball, and have at it. Replace the bread when it gets dirty, and be sure to use a variety that doesn't have seeds, nuts, or other ingredients that might scratch the paper or make a bigger mess. Soft white sandwich bread works best, although some folks swear by unseeded rye.

That's Brilliant!

Hold a wet sponge in your hand, raise your arm to wash a high surface in your bathroom, and what happens? That's right—water runs down your arm. To prevent the drips from running away, slip a terry cloth ponytail holder over your wrist before beginning. It'll neatly catch the drips before they can slide down your sleeve.

Super-Simple, Super-Safe Scouring Powder

> 1 cup of baking soda
> 1 cup of borax
> 1 cup of salt

Here's a simple cleanser that will get grease and grime off tile floors, bathroom fixtures and sinks, and just about every other surface inside—and outside—your house. It's safe, easy to use, and very effective. You'll never buy a commercial cleanser again when you try this recipe.

DIRECTIONS: Combine these ingredients, and then store the mixture in a closed container. Use it as you would any powdered cleanser when you have a tough job that no ordinary powder can solve.

Sink Spigot Cleaner 🏠 TO REMOVE SCALE from a sink spigot, fill a strong plastic bag about halfway with white vinegar, and tie the bag over the spigot so it's submerged in the vinegar. Leave it on until the crud has dissolved, then remove the bag and rinse the spigot with clear water.

Stop the Drip 🏠 IS A DRIPPING FAUCET keeping you awake all night? Halt the noise by using this old-timer's technique that does the trick on the cheap. Simply tie a piece of string around the spout, so that the end dangles into the drain. The water will flow silently down the string while you go back to bed and catch your z's. Then the only thing left for you to do is either get someone to fix the drip (or do the job yourself) and stop it for good.

G RANDMA'S OLD-TIME TIPS

Grandma Putt taught me that prevention is the key when it comes to bathroom sinks. So every month or so, when you're cleaning the bathroom, remove the drain cover and clean under its surface, clearing away hair and other gunk. While you're at it, use a bent hanger to grab hold of hair that's just out of reach.

Super Soap Scum Solution

Nasty soap scum loves setting up camp in the nooks and crannies around your tub and sink. This mixture will stake a claim on even the smallest hiding places and give gunk the boot. And the best part is that it'll deliver a super shine to all visible surfaces.

1 cup of ammonia
½ cup of baking soda
¼ cup of white vinegar
1 gal. of warm water

DIRECTIONS: Mix all of the ingredients in a bucket, stirring well. Then pull on some rubber gloves and have at it. Use a soft-bristled brush to scrub the solution into the sink and tub surfaces, and switch to an old toothbrush to dig into the cracks and crevices around the faucet and other fixtures. Make sure you wear rubber gloves as you work, and keep the area well ventilated. Rinse well.

Preventive Maintenance

SOAP SCUM IS THE BANE of many a bathroom. So keep a cellulose sponge handy, and use it to wipe out the tub after you bathe. It takes just a minute or two, but it'll cut down big-time on soap scum. Keep a squeegee or a chamois nearby, too, so that you can wipe off the walls and glass shower doors when you're finished. Then stash the tools in a wire basket that hangs over the showerhead, and they'll be ready when you are.

Ring-Around-the-Tub Scrub-a-Dub

TO REMOVE STUBBORN bathtub rings lickety-split, make a paste of 3 parts cream of tartar and 1 part hydrogen peroxide. Then rub it into the stain with an old plastic-net bath scrubber, and let it dry. Wipe off the paste with a clean, damp sponge, and your bathtub will be super-duper shiny and clean as a whistle!

Super Shortcuts

To keep soap scum from building up on your bathroom's glass shower door, just rub it down once a week with a moist cloth that's been dipped in baby oil. That's all there is to it!

Terrifically Tart Toilet Bowl Treatment

> 1 cup of borax
> ½ cup of lemon juice

Cleaning a toilet is a three-part job: the inside, the outside, and the seat. For the inside job, make stains disappear with a little citrus squeeze. This potion works like a charm, and it leaves behind a clean, lemony scent.

DIRECTIONS: Mix the ingredients, adding more borax if needed to make a paste. Wet the sides of the toilet bowl, and then rub the paste under the rim and along the surface of the bowl. Let it sit for two hours, then scrub the bowl clean with a toilet brush.

Quick Disinfectant ♠ BRIGHTEN UP A TOILET BOWL and kill germs at the same time by pouring ½ cup of chlorine bleach into the water in the bowl. Close the lid, let the bleach sit for about 10 minutes, and then scrub the stains away. **Note:** Don't mix bleach with any other cleaning products, and run the bathroom fan or open a window to exhaust the fumes during this treatment.

Kill the Germs, See the Clean ♠ IF YOU'RE CLEANING the bathroom and reach for the toilet bowl cleaner only to discover that the bottle is empty, grab some vodka instead. It'll clean your commode as well as any commercial toilet bowl cleaner—just don't waste your money on a premium brand of spirits.

PINCHING PENNIES

Toilet bowl cleaners that hang or drop into the tank seem like a great idea, but the chlorine in them can be a real troublemaker. It degrades the rubber flapper, causing that annoying (and money-wasting) trickle of water into the tank. Toilet experts call it the "vacation syndrome" because the problem happens fast when the toilet isn't flushed regularly—like when you're away on vacation. Besides, these cleaners only clean the water, so you still need to scrub the rest of the toilet. With all these drawbacks, a trusty scrub brush makes a lot more sense, and saves you dollars and cents!

Top-Notch Bowl Brightener

There's no need to pay big bucks for the fancy packaging and advertising budgets of name-brand toilet cleaners. Not when you can whip up a batch of this bowl cleaner instead. The all-natural ingredients are tough enough to do the job without doing a number on your wallet.

> 2 cups of liquid castile soap
> ¼ cup of baking soda
> ¼ cup of borax

DIRECTIONS: Mix the ingredients in a plastic container with a tight-fitting lid. You may need to add a splash or two of hot water to help blend the mixture until it is smooth. Use an old measuring cup to scoop the mixture onto the inside surfaces of the toilet bowl, and scrub it clean with a toilet brush.

H-2-Low! WHEN IT'S TIME TO CLEAN your toilet, make the job easier by lowering the water level in the bowl. Just pour about half a bucket of water all at once into the bowl. Then push your bowl brush quickly in and out of the exit hole. Finally, shut off the water to the tank at the pipe, flush, and the toilet won't be able to refill. Now you can go to town getting that bowl good and clean.

Color-Coded IF YOU REGULARLY USE yellow rubber gloves around the house, how about a purple, green, or blue pair for toilet-cleaning duty? That way, they won't get mixed up with your other gloves and spread nasty germs to other parts of the house. Store them in the bathroom, and then use them only for perking up the potty.

That's Brilliant!

To dissolve the hard-water mineral buildup in a toilet bowl, pour in a gallon of white vinegar and let it sit overnight. The acid in the vinegar will counteract the lime, loosening its grip on the once-tidy bowl. In the morning, simply scrub the mineral remains off with a nonabrasive plastic scrubbie. And if your toilet is a real tough cookie, you may have to repeat the treatment.

CHAPTER 3

LAUNDRY
Lowdown

The supermarket aisle is chock-full of detergents, stain removers, fabric softeners—you name it. That comes as no surprise because we all know how much abuse our clothing and table linens take, from grass stains and ground-in grime to red wine and ketchup spills. But with my tips and tonics, you won't have to buy commercial fabric-care products, and you won't need to spend a lot of time and effort to banish stains. You'll find everything you need to send stains packing.

All-Around Stain Zapper

Stains are one of life's little gotchas. And while each colorful blotch should generally be treated according to its individual nature, you can still tackle any stain with a basic pre-treater. So reach for this effective formula to zap smears, splotches, and blobs before you toss the dirty duds in the wash.

¼ cup of ammonia

¼ cup of white vinegar

2 tbsp. of baking soda

1 tbsp. of dishwashing liquid

1 qt. of water

DIRECTIONS: Combine all of the ingredients in a handheld sprayer bottle, put the lid on the bottle, and shake it to mix well. Keep the bottle handy near the washing machine, and reach for it whenever you need to treat a stain. Just spray the pretreater onto the spot, and work it into the fibers with an old, clean toothbrush. Wait a minute or two, and if the stain is still visible, repeat the process. Once the item is stain-free, drop it into the washer with the rest of the load, and run the machine as usual.

Out, Out, Darned Spot! 🏠 MANY STUBBORN STAINS can be removed from clothing rather easily with common household products. You can banish almost any non-oily stain with two simple ingredients: Just mix a teaspoon of white vinegar and a teaspoon of liquid detergent into 2 cups of warm water. Then use a brush or sponge to apply the mixture to the stain, and launder the garment as usual.

PINCHING PENNIES

You can save yourself a little cash and remove food stains from clothes and table linens with denture-cleaning tablets. Put the fabric in a container that's large enough to hold the stained portion, fill it with warm water, and drop in two tablets. Leave the material in the solution for the time suggested on the package, then toss the laundry in the washer. **Note:** Use denture-cleaning tablets only on color-safe fabrics.

Clothespin Cleaner

Nowadays, just about every household in the country has a clothes dryer. But I tell you, nothing beats the fresh scent of laundry that's been dried the old-fashioned way, by hanging it outdoors on a laundry line with wooden clothespins. If you're one of those folks who let the sunshine do the drying for them, you want to keep your little clipping devices as clean as the freshly washed clothes you clip them to. And with this simple recipe, you can do just that.

½ cup of bleach

1 tbsp. of laundry detergent (either dry or liquid)

2 gal. of warm water

DIRECTIONS: Mix the ingredients in a bucket, and soak the clothespins in the solution for about 10 minutes. Then pin them on the clothesline to dry in the sun. Repeat the process every couple of weeks. **Note:** Even if you use your wooden clothespins for non-laundry jobs, this routine is still a great way to fend off both dirt and mildew.

Wool Glove Saver 🏠 MAKE SURE YOUR WOOL GLOVES keep their shape after washing them. Start by shoving a wooden clothespin (either clip- or prong-type) into each finger. Then lay the gloves flat on an absorbent towel or a mesh rack to dry. That way, your gloves won't lose their shape, and will be ready to keep your hands toasty in the dead of winter when you need them most.

That's Brilliant!

Clothespins aren't used for "pinning" clothes all that much anymore, yet they're still handy little things to have around. I like to use them to hold my chip bags closed, but they also make great recipe and grocery list holders. Simply glue a magnet onto the back of a clothespin, and stick it on your fridge. This handy paper holder will allow you to consult recipes while you cook, or keep a running grocery list that the whole family can add to.

Crochet Reviver

Crocheted pieces take hours to create—and only a few seconds to get stained! If those stains refuse to yield with ordinary measures, try this magical formula to make them disappear. It gently gets rid of just about any discoloration you can think of, even if you have no idea what caused the splotch in the first place.

> 1 cup of Cascade® powdered dishwasher detergent
> 1 cup of Vivid® Ultra Liquid Bleach
> ½ cup of white vinegar

DIRECTIONS: Wearing rubber gloves to protect your skin, combine the ingredients in a large glass bowl, and submerge the stained crocheted piece into the mixture. Let it soak for about an hour, and rinse thoroughly. Dry the piece flat to prevent shrinkage, and stretch it gently to keep it in shape as it dries. If the piece is too big for the bowl, submerge only the stained part, and follow the soak with regular laundering in cold water, using Vivid according to the package directions, along with your regular laundry detergent.

Dry-Clean in the Dryer 🏠 TRY DRYEL® TO CLEAN nonwashable fabrics at home, instead of carting them to the local dry cleaner. Pretreat stains with a dab of Dryel stain remover, and then bag the items as directed with a Dryel cloth. The cloth releases cleansing moisture that loosens dirt and relaxes wrinkles. The Dryel kit contains everything you need, costs less than professional dry-cleaning services, and leaves fabrics smelling clean and fresh.

GRANDMA'S OLD-TIME TIPS

When it came to washing wool, Grandma knew exactly how to handle it. If the wool clothing was washable, she would soak it first to loosen up the dirt before starting the wash cycle. To remove dirt from felted wool, Grandma would wipe it with a dry sponge; for a more thorough cleaning, she would hold it over the steaming spout of a teakettle and brush it with a lint-free cloth.

Delicate Duds Presoak Potion

> ½ cup of hydrogen peroxide
> 4 cups of water

Soak your white or light-colored delicates in this marvelous mixture to remove stains and brighten the color. It's so gentle that you can even use it on washable silks! Just be careful with darker-colored fabric; test the potion on an inconspicuous part of the garment first to make sure the peroxide doesn't alter the color.

DIRECTIONS: Mix this solution in your kitchen or bathroom sink. Submerge your item, and let it soak for about 30 minutes. Then rinse it in clean water, and dry it on low heat, or as the care label directs.

Soft as Silk 🏠 IF YOUR SILK IS WASHABLE, use my Delicate Duds Presoak Potion (above) to remove any stains, and then wash the fabric with a protein-enriched shampoo. Silk strands are made of protein, just like hair, so the cleansing and nourishing shampoo will leave your fabric looking great. Just don't use a shampoo that includes conditioner because it'll leave the item feeling greasy.

Make Sour Smells Scram 🏠 IF YOU'VE EVER popped a load of clothes into the washer, then forgetten all about them, you know the result is sour-smelling clothes. Next time this happens, just put those duds through the wash again, but with a tablespoon or so of ammonia—no detergent. They'll come out as fresh as a field of daisies!

Q *My husband loves loading up his plate at the salad bar. Sure, it's good for his health, but it's awful for his clothes. His shirts are magnets for salad dressing. How can I get rid of these oily stains?*

A When oily salad dressings, mayonnaise, and even makeup leave their mark on your washable fabrics, start by soaking up as much of the stain as you can by covering it with cornstarch or talcum powder. Let it sit overnight, shake off the excess powder, and then treat the stain by rubbing in some liquid dishwasher detergent. Let it sit for at least 10 minutes, and then wash as usual in the hottest water the fabric can stand.

Doggone-Good Fabric Freshener

> 2 cups of baking soda
> 2 cups of unscented liquid fabric softener
> 4 cups of warm water
> Scented oil

If your upholstered furniture is beginning to smell just like your favorite four-legged friend, you need to take action. Deodorizing sprays don't zap the stink for very long, so use this nifty, thrifty spritzer instead.

DIRECTIONS: Mix the baking soda and fabric softener in the water, and add a few drops of your favorite essential oil, such as lemon, orange, or almond (let your nose be your guide as to quantity). Pour the mixture into a handheld sprayer bottle, and use it on upholstered furniture, draperies, carpets, or any other fabric that smells less than springtime fresh. **Note:** Test this on an inconspicuous area first to make sure it doesn't stain the fabric.

It'll All Come Out in the Wash 🏠 YOUR WASHING MACHINE pretty much cleans itself on the inside, but that doesn't mean you're home free: Liquid fabric softeners, which are very waxy, can build up in the washer's innards and make a real mess. You can avoid the whole ball of wax in two ways: If your machine has a fabric-softener dispenser, pour in the recommended amount of softener and then add about ⅓ cup of water. This will help the softener dissolve properly. For machines without dispensers, mix ⅔ cup of softener with ⅓ cup of water in a bowl or cup, and pour it into the washer tub.

That's Brilliant!

Not only can fabric-softener sheets make your clean laundry soft and fresh, but they can also help your waiting-to-get-washed laundry smell a bit better. When you pull the used sheet out of the dryer, throw it right into the hamper. That way, you won't have to be chased out of your own laundry room by your teenager's smelly socks, hubby's sweaty gym clothes, or the pile of underwear that your toddler decided to hide in the back of the closet two months ago!

Down-Home Dry-Cleaning Formula

It's great to take your clothes to the dry cleaner, because you know they'll come back clean as a whistle. But dry cleaning can get pretty pricey. So if you have yucca or soapwort (*Saponaria officinalis*) growing in your garden, you're in luck! You've got the fixings for a super-gentle "dry-cleaning" solution for delicate fabrics, woven baskets, and other nonwashable items.

DIRECTIONS: Put the powdered yucca or soapwort root in a blender and add the boiling water. Cover, and blend on high for two to three minutes. Allow the mixture to settle for a few seconds before removing the lid. Scoop out the suds with your hand, a cloth, or a soft brush, and rub them into the item you want to clean. Let the foam dry, then simply brush it off.

The Sauna Treatment

WASHABLE WOOLS are a real miracle fabric, but frequent washing will cause the fibers to break down before their time, especially if you use a top-loading machine that's got an agitator in the middle. So instead of machine washing wool, freshen it by hanging it up in a steamy bathroom. The moisture will lift the fibers, fluffing up the surface and easing out any wrinkles. This is a great trick to keep up your sleeve when you're traveling.

Super Shortcuts

Warm fleece and fake fur made from synthetics are usually acrylic or modacrylic, so they're amazingly easy (and cheap) to care for. Most are machine washable on the gentle cycle and will do just fine in a dryer on low heat. The only drawback to these fab fakes is that they tend to build up a lot of static electricity, making them clingy or even . . . shocking! The solution? Just pop a fabric-softener sheet (or two) into the dryer to reduce the snap, crackle, and pop.

Dryer Lint Modeling Clay

1 ½ cups of dryer lint
1 cup of water
½ cup of all-purpose flour
Food coloring (optional)

Running out of rainy-day activities for your cooped-up kiddos? Is regular clay a little too messy or expensive for your budget? Young (and not-so-young) arts-and-crafters can have a ball with this trash-to-treasure art medium—both making it and using it.

DIRECTIONS: Put the lint in a saucepan, cover it with the water, and let it sit until the lint is saturated. Add the flour, and stir until the mixture is smooth. Add 2 to 3 drops of food coloring, if you like. Cook the mixture over low heat, stirring constantly until it holds together and you can form peaks with the spoon. Pour the mixture onto a cutting board, sheet of aluminum foil, or stack of newspapers, and let the clay sit until it has cooled. Then let your creative juices flow. When you've finished your masterpiece, let it dry for three to five days. After that, you can leave it unembellished, paint it, or deck it out with the trimmings of your choice.

Fire-Starter Nuggets 🏠 YOU CAN USE your dryer lint to make sure you always have an easy way to start your fire (and impress your friends in the process). First, cut a cardboard egg carton into 12 sections, and fill each one with dryer lint (but only from 100 percent cotton fabric). Then melt down old candle stubs or paraffin sealers from homemade jelly, and pour a layer of melted wax on top of the lint. When it's time to light your fireplace or charcoal grill, set one of your nuggets in the kindling or briquettes, and hold a match to the cardboard edge. You'll have a roaring fire in no time.

That's Brilliant!

Toss 100 percent cotton lint onto your compost pile or into your compost bin. Or if you have no place to "cook" compost, simply bury the lint among your plants. It'll break down in a flash, adding valuable organic matter to the soil.

Fragile-Fabric Soap

Instead of buying special liquid soap to clean your delicate fabrics, mix up a batch of this remarkable recipe. It keeps practically forever without losing its gentle cleaning power, so it'll be ready whenever you need to handle your hand washables with care!

> ½ cup of borax
> ½ cup of powdered laundry detergent
> 2 cups of water

DIRECTIONS: Combine the ingredients in a saucepan, and simmer the mixture over low heat for about 10 minutes, stirring constantly. Beat it with a wire whisk occasionally to break up the lumps. Let it cool, and then pour it into a clean plastic bottle. Use it just as you would any commercial product for delicate fabrics.

Nylon News 🔺 THE ORIGINAL MIRACLE FABRIC—nylon—is strong, lightweight, wrinkle-resistant, and easy to care for. Just pop your nylon items in the washing machine, then spin them dry on low heat—it doesn't take long! Nylon tends to shrug off stains because it's not very absorbent, but it does hoard static electricity. So add a fabric-softener sheet to the dryer to take care of that clingy little problem, and your nylon will hang smoothly once it's dry.

Save the Spandex 🔺 KEEP YOUR SPANDEX STRETCHY—without stretching it out—by avoiding chlorine bleach, which can damage the material; use an oxygen bleach instead. Heat can be harmful, too, so always dry garments made of spandex or spandex blends on low heat, and take them out of the dryer just as soon as they're dry. Or better yet, air-dry them instead.

PINCHING PENNIES

If you're willing to risk it, you may be able to hand wash certain dry-clean-only items and save yourself some cash. White or light-colored fabrics and unstructured items like sofa throws, sweaters, and lingerie hold up best to hand washing. Bright colors, dark and black hues, and tailored clothes like blazers and dress slacks usually don't.

Freshen-Up Fabric Spray

½ cup of baking soda
½ cup of fabric softener
1 cup of warm water

Those fabric-deodorizing sprays that are sold at the supermarket are a fine invention, and they're all the rage these days. It's no wonder: They help keep your clothes and furniture smelling fresh and clean, even when they're truly "lived in." But, if you use these sprays often, the cost can really add up. Instead, make your own with this handy recipe.

DIRECTIONS: Mix all of the ingredients in a bucket, then pour the solution into a handheld sprayer bottle. Now you're ready to spray the deodorizer in gym bags and on pet beds, sofa cushions, or anywhere that can use a little freshening up.

Wrinkle Reducer

COTTON AND LINEN will shrink and wrinkle unless they're specially treated. So be sure to follow the instructions on the care label. If the label's missing, play it safe and launder them in cold water and dry them on the cool setting to cut down on shrinkage. And never let them sit in the dryer once they're done—unless you really love to iron! But if you do forget to take cotton or linen items out of the dryer at the end of the cycle, try this neat trick for getting rid of the worst of the wrinkles: Just wet a clean white cloth, toss it in with the wrinkled load, and run the dryer again. The moisture will relax the wrinkles. And this time, when the cloth is dry, remove the clothes immediately so you don't have to repeat the rescue process again.

Super Shortcuts

Fluffy acrylic, modacrylic, and acrylic blends are much easier to clean than the wool they are meant to resemble—you can usually toss these synthetic fabrics in the washer and dryer without worry. But check each label anyway because some items require hand washing or dry cleaning. Use low heat in the dryer, and don't forget to empty the lint trap afterward because acrylic fabrics tend to shed fibers while they tumble around.

Glue Spot Be Gone Solution

Keep this mixture on hand for whenever you get clear household glue on finicky fabrics—if they're washable—including rayon, acetate, triacetate, wool, and silk. And if the label says "dry-clean only," spend the extra money and let a pro deal with the stain.

> **1 part dishwashing liquid**
> **1 part glycerin**
> **8 parts water**
> **1 tsp. of white vinegar**

DIRECTIONS: Mix the ingredients in a plastic squeeze bottle, and shake it well before you use it. To remove glue stains from those special fabrics, squeeze the potion onto a cloth and use it to wipe the fabric, working outward from the center of the spot. If any glue remains, cover the stain with a paper towel that's saturated in the potion, let it sit for about 15 minutes, and wipe the remains away. Then rinse the fabric with clean water, and launder the item with an enzyme detergent.

Don't Feed the Pests 🏠 FOOD, BABY FORMULA, perspiration, and urine stains are all manna from heaven to clothes moths. They'll zero in on any wool, silk, and other animal-based fabrics that have these kinds of stains, and they'll even attack cotton, linen, and synthetic fabrics if the stains are edible. So the moral of the story is: Never hang up dirty clothes in a closet, or store them away for the season, without getting those stains out first!

GRANDMA'S OLD-TIME TIPS

Crayon, chocolate, spaghetti sauce, lipstick, and the like are combination stains—they leave both grease and a dye behind. So when you get dark, greasy gravy or a blob of pizza sauce on your clothes, do what Grandma Putt did. She tackled the stains in stages, treating the grease first (see FAQ on page 75). That treatment sometimes made the dye disappear, too. But, if it didn't, she would then dab on some ammonia or bleach that was safe for the fabric, and the stain simply washed away.

Hair's the Fabric Softener

This fabric softener uses ingredients you always have on hand. It'll make your laundry soft and smelling as clean as freshly shampooed hair. It's easy enough to add it to your washing machine's rinse cycle, but if you miss the start of the cycle, no worries. You can take care of the job in the dryer, instead.

> 1 cup of white vinegar
> ¾ cup of any brand hair conditioner
> 2 cups of water

DIRECTIONS: Stir the ingredients together in a bowl, and pour the mixture into a clean bottle with a tight-fitting lid. Add ¼ cup of the softener to your washing machine's rinse cycle, or sparingly spray some of the liquid onto a washcloth and toss it into the dryer along with your freshly laundered clothes.

Fabric-Softener Hair Spray 🏠 FABRIC SOFTENER makes a fine stand-in for hair spray. Mix 1 part softener and 2 or 3 parts water (depending on how firm of a hold you want) in a handheld sprayer bottle, and spritz the solution onto your styled hair. It'll keep those locks in place and make them shine like the sun!

Stop the Sock Monster 🏠 PREVENT SOCK LOSS by binding each pair together with a safety pin before you toss them into the washing machine.

Super Shortcuts

✂ Fresh out of hair conditioner? No problem! Just mosey on into the laundry room, and pick up the liquid fabric softener. Use it as you would your normal conditioner, pouring a similar-sized dollop of the liquid into your hand and working it through your hair. Wait two or three minutes, and rinse. **Note:** Be sure you use a quality product for this job. The lower-priced fabric-softener brands don't seem to work as well—or so I'm told by my lady friends, who have a lot more hair than I have!

Homemade Dryer Sheets

> 1 part liquid
> fabric softener
> 1 part water
> Washcloth

Dryer sheets are just the ticket for keeping your clothes soft and static-free. But what a waste to spend money on something you toss after only one use, when you can make a dryer sheet yourself and use it over and over again!

DIRECTIONS: In a medium-sized mixing bowl, mix the fabric softener and water together thoroughly. Soak the washcloth in the mixture for a minute or two. Wring it out, and toss it in your dryer along with a load of wet laundry to prevent static cling. You can reuse the washcloth several times before laundering it and starting with a new batch of softener.

Vim and Vinegar 🧺 THE TELLTALE STENCH that's coming from your laundry room can only mean one thing: You forgot the load of laundry in your washing machine, and now the wet clothes smell mildewy. Don't toss 'em in the dryer yet—even a dryer sheet won't be able to hide the stink. Instead, run 'em through the washer again. But this time, pour 2 cups of white vinegar in, and set the machine to a hot-water wash cycle. Then, wash them again with detergent before drying.

Holey Hoses! 🧺 WASHING MACHINE HOSES have about a five-year life span, but they can go at any time. A sudden rupture will spew water everywhere, and you don't want that to happen! So check your hoses once a month. If you see a small blister or any cracking, buy a new hose, pronto!

That's Brilliant!

When the enamel coating inside a washing machine or dryer drum gets chipped, nasty rust spots may form on the bare metal—which will leave streaks and stains on your clothes. So buy a touch-up kit at a local hardware store now to fix the dings as soon as you notice them. Dab some touch-up paint on any chips around the machine doors, too, because rust can get started when you transfer wet clothes.

Intensive Care Wall Wipe-Up

> 2 oz. of borax
> 1 tsp. of ammonia
> 1 bucket of water

You've just purchased a new dryer. Now you need to get rid of the old one—and it's been in the same spot since 1966! You could probably knit a sweater with all the lint underneath it. Lint, dirt, and dust can be easily swept away, but stains on the walls behind old appliances are more of a challenge. Use this solution to wash grime from your laundry room walls.

DIRECTIONS: Dissolve the borax in the ammonia and water, dip a sponge into the solution, and scrub down the wall, starting at the top. Dry the wall with a clean, soft cloth or paper towel immediately. **Note:** If you live in a home that was painted by the builder, only a thin layer of paint may have been sprayed on the walls. Wash a small section as a test to be sure the paint doesn't come off.

Bust the Dust 🏠 LINT DUST DOESN'T build up on laundry room walls as quickly as it does on tables and other horizontal surfaces, so you'll need to dust them only once every few months. Use a long-handled duster or the upholstery brush attachment on a vacuum cleaner to do the job, and work from the top down. Every week or so, sweep away lint in the corners, where the walls meet the floor and ceiling.

Q *A friend of mine had a fire in her dryer, caused by a buildup of lint. She had to replace the appliance, and it took weeks for the smoky smell to dissipate. How can I make sure my dryer won't do the same thing?*

A Lint is a dryer's worst enemy, and as your friend found out, it can be yours, too. To keep your dryer working—and a roof over your head—here's what you can do. After each load of laundry, clean out the lint screen. Then, every month or so, wash the lint screen in warm, soapy water. Let it air-dry overnight, and come morning, it'll be ready for action. About every four months or so, scrape built-up lint from the outdoor vent using a stiff-bristled brush. Finally, twice a year, remove the exhaust duct from the dryer, and give it the once-over with a vacuum cleaner.

Laundry Detergent Supercharger

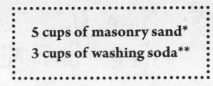

5 cups of masonry sand*
3 cups of washing soda**

Laundry detergent is a great invention. It allows laundry to get a hundred times cleaner than our forefathers (and mothers!) ever could have imagined. But even with all of the advances in laundry soap technology, detergent alone sometimes needs a little extra boost. This concoction has just what you need to get your laundry super clean.

DIRECTIONS: Put the sand and soda together in a 1-gallon resealable container. Shake the mixture vigorously to blend it. Then add 1 to 2 tablespoons of the supercharged solution to each laundry load to give your detergent more power.

*Available at home-improvement stores.
**Available online and at some hardware stores.

Scour with Powder 🏠 POWDERED LAUNDRY DETERGENT makes a fine scouring powder for kitchen and bathroom sinks, tubs, and other ceramic or porcelain surfaces. Sprinkle some dry detergent into the sink or tub, moisten with water, and use a nonabrasive plastic scrubbie to go to town.

Clean Off the Grime 🏠 WHEN MESSY OUTDOOR CHORES leave you with greasy, grimy hands, wash them with laundry detergent (either liquid or dry). Those mitts'll be as clean as a whistle in no time!

Makeshift Laundry Bag 🏠 MAKE A LAUNDRY BAG by running a drawstring through the top of a pillowcase. In fact, if you know a youngster who's headed off to college, make three bags—color-coded for what Grandma called "the patriotic sorting method." That way (with luck), the young scholar will remember to keep her red shirts, white socks, and blue jeans separate in the wash.

PINCHING PENNIES

Need a new laundry basket? Don't go out and buy one. Instead, rustle up that old baby bassinet and assign it to washday duty.

Mosey On Garden Stain Mixer

White non-gel toothpaste
1 tsp. of hydrogen peroxide
Water

Spending the day gardening can leave you with a great sense of accomplishment. But it can also leave you with some pretty interesting stains. For those times when you come in from the garden wearing, well, a bit too much of the great outdoors, try this old-time routine.

DIRECTIONS: Mix a dab of toothpaste with the hydrogen peroxide. Rub the paste onto the spot with a clean, soft cloth, and rinse with clear water. Then toss the duds into the wash so they're ready for your next gardening go-round.

Say Farewell to the Smell 🏠 Laundry detergents clean your clothes, but they don't always wash away stubborn, lingering odors, like those from cigarette or campfire smoke. Scented detergents simply mask persistent aromas without really getting rid of the stink—and, let's face it, those cloying perfumes aren't for everyone. So give your next stinky load of laundry a deodorizing boost by adding ½ cup of white vinegar to the wash cycle, along with your regular detergent.

GRANDMA'S OLD-TIME TIPS

My Grandma Putt shaved a piece of Fels-Naptha® soap into every load of her laundry, but when she needed extra power to eliminate stains, she'd rub the bar right on the cloth like a stain stick. Fels-Naptha works like magic on grass stains, chocolate stains, and all kinds of other spots. When I was a kid, we even scrubbed it onto our skin to prevent poison ivy! To put this old-fashioned helper to work on your grass stains, wet the stained fabric and vigorously rub the soap into the spot before you launder the item as usual. Look for Fels-Naptha in the laundry section of your supermarket.

Rust-Removing Laundry Paste

Salt

White vinegar

4 cups of boiling water

It sure can be aggravating to discover mysterious rust stains on what should be clean laundry. After all, you've just run the clothes through the washing machine. If the drum of your washer appears to be rust-free, check out your bleach dispenser. Sometimes rust forms under the funnel, and you can't see it unless the funnel is removed. Now, here's a super solution that'll bust the rust.

DIRECTIONS: Mix the salt and vinegar together until a thin paste is formed. Thickly spread the paste over the rusty spots, and let it sit for 30 minutes or so. Lay the fabric over a large pot, and pour the boiling water through the fabric, right through the spot where the marks are. Allow the fabric to dry, then check the stain. Repeat the treatment if you see any signs of rust. Once the stain is gone, run the garment through a complete wash cycle.

What a Rhubarb! YOU MIGHT FIND THIS a little strange, but believe me, it's true—rhubarb juice is another great solution to use when your clothes or other washable items get stained with rust. Start by cutting five stalks of rhubarb into half-inch pieces, and then simmer them in 2 cups of water until they're soft. Strain off the rhubarb juice into a heat-proof bowl, and submerge the stained area of the garment in the hot liquid until the stains disappear. Launder the item as usual, and the rust will be history.

That's Brilliant!

This tried-and-true rust remover has been used for decades by military personnel on metal, floors, upholstery, and even clothes. Just rub a dab of naval jelly into the stain, and let it sit for about 10 minutes. Wash it away with soapy water, or launder the item as usual. Naval jelly is actually a form of phosphoric acid, so handle it with care, and make sure your skin and eyes are well protected. You'll find this product at most hardware stores.

Set-In Stain Solution

If there's one thing that clothes attract more than anything else (including attention), it's stains. Sometimes, those nasty spots get so worked in before we notice them that no amount of washing will get them out. But don't give up hope. Give this solution a shot, and say "so long" to set-in stains.

½ cup of chlorine bleach

½ cup of powdered dishwasher detergent

2 gal. of hot water

DIRECTIONS: Mix all of the ingredients in a large bucket. Then soak your clothes according to their type: White cottons and washable synthetics should be soaked for two hours (or more, if they're heavily stained). For colored clothes that can be bleached, allow the mixture to cool and then soak them for 30 minutes. Once the time is up, remove the garments from the bucket, and launder as usual. **Note:** Read your garments' care labels carefully before using this solution. Not all colors can be bleached, and you don't want to end up with bleached-out patches all over your favorite clothes.

Color Keeper 🏠 THE NEXT TIME you come home from a trip to the clothing store, grab the vinegar on the way to the laundry room. Add ½ cup of white vinegar to the wash cycle the first time you launder any new clothes. Not only will it help clean the manufacturer's smelly chemicals out of the fabric, but the vinegar will also set the colors, so they'll stay bright longer.

Q *Baseball and barbecue season means chomping down on juicy, plump hot dogs with plenty of mustard. But I always end up with mustard on my clothes. What's your trick for getting these stains out?*

A You're really speaking my language—I love a good ball game, *and* a good hot dog! And I'm not known for skimping on the mustard, either. When your hot dog goes haywire, make quick work of mustard stains by scrubbing them with a solution of 2 parts water and 1 part rubbing alcohol, and then wash your clothes as usual.

Smoke-Smell Banishing Rinse

½ cup of baking soda
Laundry detergent
1 cup of white vinegar

So you spent a nice evening singing and roasting marshmallows around the fire pit. But now everything you wore smells like wood smoke. Never fear, a smoke-busting solution is here!

DIRECTIONS: To rid your washable duds of the smoky smell, add the baking soda to your laundry along with your regular detergent during the wash cycle, and then finish the treatment by pouring in the vinegar at the start of the rinse cycle. These tricks work great for cigarette smoke, too, so if you're a smoker, no one will be able to smell you coming! For woolly sweaters and other nonwashables, just hang them outside on a breezy day and let the smoky smell blow away.

No Smoking Zone! 🏠 IF YOUR WASHING MACHINE smells smoky (and you know nothing's burning), I'll bet my bottom soap flake that you've put either too many clothes or too little water in the tub. That burning smell is probably coming from an overstressed drive belt. So stop the machine, make the appropriate adjustments to the water level or clothing amount, and then restart. And pay attention to the settings before you start a load so it won't happen again!

Super Shortcuts

✂ Take a minute to read the care label in your clothes and on other items before you clean them—or, better yet, before you buy them! That's the easiest way to figure out exactly what kind of TLC they'll need. Not only will you learn whether that blouse is washable, but you'll also find out what cycle to use in the washer and dryer, how to iron it, and whether or not you can use bleach on it. With those basics in place, the only thing you'll need is advice on how to make stains disappear—and that's why I'm here!—to help you keep your fabrics in tip-top shape.

Super-Softening Laundry Sachet

½ cup of baking soda
2 tbsp. of cornstarch
2 drops of your favorite scented oil

Get ready for super-soft laundry when you make this laundry sachet. It softens load after load, and leaves your clothes, sheets, and bath towels smelling fresher than any flimsy fabric-softener sheet ever could.

DIRECTIONS: Stir the baking soda and cornstarch together in a bowl. Now find an orphaned sock with tightly knit fibers. Fill the toe of the sock with the powdery mixture, and drip in the essential oil. Tie the sock shut as tightly as you can. Toss it into your dryer along with a load of freshly washed laundry. Your items will come out of the dryer feeling soft and smelling fresh and clean. You can reuse the sachet as many times as you like until the fragrance fades. Then simply untie the sock, empty the sachet, and refill it with a fresh batch.

Are You Gellin'? 🏠 MAKING A BATCH of homemade fabric softener to add during your washing machine's rinse cycle is super easy and costs only pennies. Not only that, but it'll also leave your laundry cuddly soft. Simply add 1 tablespoon of unflavored gelatin to 1½ cups of boiling water and stir it until the gelatin is completely dissolved. That's all there is to it!

Q *A friend of mine insists that it does no harm to do laundry with as little water as possible. I'm not so sure. Just how much water should I be putting in the washer to get my clothes clean?*

A This advice might seem odd coming from an old tightwad like me, but trust me on this: When you toss a load of laundry into a top-loading machine, always use more water than you think you'll need. If there's not enough water in the tub, the clothes can't circulate freely, and they'll get all wrapped around the agitator. That'll cause wear and tear on the agitator seal—not to mention the damage all that twisting and turning will do to your clothes.

Tender Lovin' Tonic for Treasured Quilts

Buttermilk (with butterfat content of 1% or less)

Lemon juice

Heirloom quilts and other old textiles need special TLC. Any quilt that can be considered an heirloom is probably a pretty important part of your household because of the history and cherished memories attached to it. Treat treasured quilts oh-so-tenderly with this gentle tonic.

DIRECTIONS: First, test the fabric for colorfastness by rubbing a damp white cloth on each different color of fabric and thread. (Don't just assume that because the red in one print didn't bleed, the red in another one won't—the dyes could be quite different.) Then, fill your washing machine with cold water, and for every gallon, add 1 quart of buttermilk and 1 tablespoon of lemon juice. When the tub is full, stop the machine, and gently hand-agitate your treasure. Rinse the quilt thoroughly, and hang it up to dry.

Blowin' in the Wind

WHETHER A HAND-MADE QUILT is old or new, the best way to get it clean is to simply air it out on a clear, breezy day. Let the quilt blow in the breeze for a few hours to shake off the dust and leave it with a fresh scent. And in case you're wondering, dry cleaning is a no-no for old quilts—the chemicals might damage the fabric or thread.

Super Shortcuts

If you have a quilt or piece of clothing made from vintage fabrics, don't try to save time by machine washing the item, or it may fall apart! Instead, start with a simple soak in cool water and mild soap. Swish the item around in the soapy water, but don't rub. Let it soak for at least an hour, and rinse multiple times in cool water until the water runs clear. And don't wring out the item—squeeze it very gently, then roll it up in thick towels and pat to get the water out. Then lay it flat on a dry towel, away from direct sunlight.

Wool Sweater Wash

Without a little TLC, wool can become lumpy, bumpy, pilly, and downright scratchy. Whether you get it dry-cleaned or use a special at-home cleaning product, wool fabric requires special care. You wouldn't want your wool to become too itchy, would you? So give this gentle formula a try.

DIRECTIONS: Dissolve the dishwashing liquid in a sink filled with the lukewarm water. Turn your wool garment inside out, and submerge it in the water for three to five minutes. Squeeze out the suds and soil (don't wring or twist), then rinse the garment two or three times in clean water. Once all the suds are gone, lay the item flat to dry.

Keep Wool in Excellent Shape 🏠 WOOL GARMENTS NEED to be kept looking their best. So take a few minutes to care for them with these easy tips, and you'll be rewarded with garments that will look good as new for years to come: Just hang wool garments on padded or shaped hangers so the shoulders don't get misshapen. Before you hang them up, make sure the pockets are empty, and all buttons and zippers are closed. Also, get all nonwashable wools dry-cleaned before you store them, and make sure they're folded when they go in drawers.

That's Brilliant!

Whoops! You (or a well-meaning helper) accidentally sent your best wool sweater through the washing machine, and now it's a shapeless lump. Is it ruined? Maybe. But before you give up hope, soak the sweater in tepid water with a squirt or two of good-quality shampoo added to it. This may soften the fibers enough to let you reshape the garment. Stretch it out and shape it as best you can, then lay it flat to dry. With any luck, your sweater will be back in wearable shape just in time for sweater weather.

CHAPTER 4

ROOM
for Improvement

What makes a house a home? My money's on love and kindness—and dirty baseboards, grimy windows, and sour smells won't change that. But even the best of homes can get a bit "tired" from time to time. If that sounds like your humble abode, don't despair. Whipping it into shape doesn't take an army of volunteers—though enlisting the help of your loved ones for a bit of cleaning wouldn't hurt! So put on your work clothes and follow my advice for a house you can be proud of.

Awfully Easy Artist's Paint

Rainy days are meant for sitting with your kids as they paint a masterpiece. But don't worry if you're fresh out of craft paint the next time storm clouds gather. Encourage your young artists' ventures by giving them a steady supply of this easy-to-make paint whenever their creative spark shows itself.

½ cup of soap powder (such as Ivory Snow®)

6 cups of water

1 cup of liquid laundry starch

Food coloring

DIRECTIONS: Dissolve the soap powder in the water. Mix in the starch, and add the food coloring of your choice. (The amount is up to you—the more you use, the darker the shade will be.) Store any leftover paint in a container with a tight-fitting lid until the next rainy day.

Pour on the Alcohol 🏠 DON'T THROW PAINT-STAINED clothes into the ragbag when you can make those spots vanish in a hurry. Place the stained part on top of an old towel, pour rubbing alcohol on the spots, and the paint will dissolve like magic. Help it along by rubbing the fabric with an old toothbrush to get out deep-down stains, and then launder the item as usual. This trick works great on most clothes, but if you're not sure how the fabric will handle it, test it in a hidden spot first.

PINCHING PENNIES

Got a little paint on your sneakers? Well, don't toss them in the trash. Give this simple solution a try the next time you splatter paint on your sneakers. First, spray the spots with hair spray, and then scrub the stains away with a nonsoapy steel wool pad. Don't rub too hard—a light touch is all you need to make the paint disappear. This method also cleans paint off old leather work boots, but don't use it on high-fashion shoes because the steel wool will scratch the surface.

Baseboard Bonanza

Don't ignore dust-accumulating baseboards when scrubbing your floors. It's easy to forget that they need to be spruced up just as often as anything else. Cleaning them is simpler if you do it regularly—instead of letting grime accumulate. Start by vacuuming them with a hose attachment, then wipe 'em down with this DIY spray.

> 1 tbsp. of cornstarch
> 2 cups of boiling water
> ⅓ cup of white vinegar

DIRECTIONS: Measure the cornstarch into a handheld sprayer bottle, then carefully add the water and stir until the cornstarch has dissolved. Add the vinegar and stir. Now take aim and fire the spray on baseboards, wiping them clean with a soft cloth or damp sponge.

An Upright Solution

EVEN IF YOU DON'T USE a sponge mop to clean your floors, pick one up the next time you're out and about, and use it to spiff up your baseboards. Mix a solution of ½ cup of vinegar and 3 cups of warm water in a bucket, dip in the mop, and wipe down your baseboards. That's all you'll need to clean away the pet hair, dust, and other crud without backbreaking, knee-crushing effort. In fact, you'll barely have to bend over to get the job done!

That's Brilliant!

Electric hair dryers have been around since the roaring twenties. But not until the swingin' sixties did the familiar pistol-grip blow-dryer become a fixture in bathrooms all across the country. And these gadgets can do a lot more than dry hair. Select the coolest setting on your blow-dryer, take aim, and use the thing to blow dust to Kingdom Come! This trick is just the ticket to tackle dust on baseboards, lamp shades, fabric hangings, artificial flowers, and anything with intricate carving or relief work—like wooden furniture or ceramic pieces. You can even blow dust out from behind radiators, bookcases, and refrigerators!

Bravo Brass Brightener

My Grandma Putt was button-popping proud of her brass hardware, so she always made sure that it looked bright and shiny. Grandma cleaned all of the brass in her house with this miraculous mixture, which also works like a dream on copper cookware.

DIRECTIONS: Mix the flour, detergent, and salt in a bowl. Add the remaining ingredients, and blend thoroughly. Dip a soft, clean cotton cloth into the mixture and rub it onto the brass, taking care to get it into all the nooks and crannies. Buff with a second clean cloth. Store any leftover cleaner in a jar with a tight-fitting lid.

> ½ cup of all-purpose flour
> ½ cup of powdered laundry detergent (without bleach)
> ½ cup of salt
> ¾ cup of white vinegar
> ¼ cup of lemon juice (fresh or bottled)
> ½ cup of hot water

Top Brass 🏠 A MELLOW GLOW IS part of the charm of antique brass, so don't scrub away the patina along with the crud. Instead of using an abrasive cleaner, give your brass a bath and a buff. Here's how: First, wash the object in hot, soapy water to get off the grime and any old wax. Rinse and dry the piece thoroughly, and then remove any deeper dirt by rubbing the brass with a soft cloth moistened with boiled linseed oil. When the brass is nice and clean, buff it with a soft, dry cloth.

Lick It with Lemon 🏠 TO QUICKLY CLEAN GRUNGY BRASS, cut a lemon in half, sprinkle it with baking soda, and rub the lemon directly on the brass. Then wipe the residue off with a clean, damp cloth, and buff the brass to a beautiful shine.

PINCHING PENNIES

Always keep the word *gentle* on your mind when you're cleaning brass because it's all too easy to scratch the soft surface of this coppery metal. So put away that harsh scouring powder, and give mildly abrasive (and free!) wood ashes a try. Simply sprinkle a soft cloth with powdery wood ashes—no charcoal chunks or splinters, please!—and add a little elbow grease to make that dullness disappear in a flash.

Clean Slate Solution

Slate is a terrific, durable surface to use in your kitchen or bathroom, as flooring, or around your fireplace. To keep slate looking clean and stain-free, mix up a batch of this "eraser" with ingredients from the kitchen.

DIRECTIONS: Make a thin paste by adding baking soda to equal parts of lemon juice and water. Gently rub the stain with a soft cloth that's been dipped in the paste. Then rinse the surface with clear water and let it dry.

The Wet Look

UNSEALED SLATE TURNS dull and pale as it dries. So if you like the wet look, you need to apply a glossy acrylic slate sealant to the slab. Follow the directions on the label, and give it a couple of coats for better protection. Use a paint pad to apply the sealant; its rectangular shape will allow you to get into corners and along the edges without glopping any sealant on the baseboards or walls.

Tell the Rubber to Hit the Road!

RUBBER MATS and non-skid rubber rug pads are big no-nos on a slate floor. They stick to the slate and make an awful mess. If this happens to you, moisten the residue with mineral spirits, allow it to soak for several minutes, then use a plastic putty knife to gently remove as much gunk as you can. Dampen a soft cloth with more mineral spirits and wipe the remaining spot. When it's gone, soft-mop using 1 cup of mild laundry detergent in a gallon of water.

FAQ?

Q My slate countertop has lots of scratches from slicing bread and gosh knows what else. Is there any way to remove the marks without calling in the pros?

A You betcha! Just buy a bottle of food-grade mineral oil from your local pharmacy, and rub it in all over the countertop. Then perform this test: Run a knife lightly across the slate to make a scratch about an inch or two long. Then rub the scratch with your fingertip—the oil in the slate should make the mark disappear like magic. Reapply the oil once or twice a month, and your countertop will be a real smoothie.

Clear the Air Potpourri

There's nothing quite as welcoming as a house that's filled with a delightful aroma. Keep your home smelling terrific with this must-busting, air-clearing potpourri.

DIRECTIONS: Blend all of the dry ingredients, then scoop a cup of this herbal mixture into a pot along with the water. Let the brew simmer until the fragrance permeates the whole house. Store the rest of the mixture in an airtight container in a cool, dry place so it's ready to clear the air whenever you need it.

> 2 cups of dried thuja cedar twigs and leaves
>
> 1 cup each of hyssop flowers, calamus root, and elecampane root
>
> 4 cups of water

Put the Kettle On 🏠 TO ADD A LITTLE FRAGRANCE to the air, put a large pot of water on the stove to boil, and add some cinnamon sticks and cloves to it. Let the contents simmer until the water's nearly gone. Then you can either put the whole shebang away, or just add more water and keep on boiling it. I especially like to do this in the winter when the air is dry. Not only do the cinnamon sticks and cloves make the air smell wonderful, but the simmering water also adds some much-needed humidity to the room. Another bonus—if your house is on the market, this little trick will help give potential buyers positive vibes.

PINCHING PENNIES

You've just found an old trunk in the attic, and it's full of books. But boy, oh boy, does it ever smell musty. Here's how to get rid of the odor and save the books. All you have to do is put a few charcoal briquettes, a scoop or two of clay cat litter, or a sprinkling of baking soda inside the trunk and close it up. Keep changing the odor absorber every few days until the musty smell is gone. As for the stinky books, put them inside a paper bag for a few days with the charcoal, litter, or baking soda.

Cracked-Ceiling Camouflage Paint

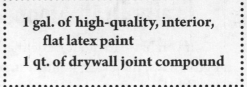

1 gal. of high-quality, interior, flat latex paint

1 qt. of drywall joint compound

If your basement (or any room in your house for that matter) has a plaster ceiling, it's bound to pick up a few imperfections over the years. In order to hide those cracks or holes, touch it up with this paint. It'll give the ceiling a slightly irregular surface and fill in any small cracks.

DIRECTIONS: Pour the paint into a bucket (at least 2 gallons), and then add the joint compound. Mix them together with a stir stick. Dip in a roller and apply two coats to your ceiling. You'll get it looking like new again in no time.

Mix Paint Quickly 🏠 IT TAKES A heck of a lot of stirring to get your paint fully mixed. But you can mix it in a flash using a paddle attachment on your electric drill. In order to prevent spattering, place the can in a paper bag before mixing the paint. That way, when you're done mixing, you can just pull out the can of paint and throw away the bag.

Ceiling Helper 🏠 THE OLD HOLLOW rubber ball that you found in the basement may have lost its bounce, but it can come in mighty handy if you're planning to brush-paint a ceiling or overhead woodwork. Just slice the sphere in half, cut a slit in one of the halves, and slide it onto your paintbrush handle, hollow side up. It'll catch the drips that fall from above, leaving you with paint-free hair and arms.

Q *I have an old shower curtain that I just don't know what to do with. There's truly nothing wrong with it, so there has to be a new use for it, right?*

FAQ ?

A Right you are. For one, old vinyl shower curtains make great drop cloths. Simply spread them out on floors and over furniture to protect them from wayward drips.

Crafty Draft Stopper

If you have an old necktie that's seen better days, then turn it into a snazzy draft blocker for your gusty front or back door! All you need are a few odds and ends, and one drafty door.

> 1 old or unwanted necktie
> Thread or fabric glue
> Stuffing, such as batting or old panty hose
> Dry beans

DIRECTIONS: Fold the tie in half lengthwise, and either sew or glue the back closed. Then sew the narrow end together to form an open-ended tube shape. Loosely insert your stuffing, and then add the dry beans to give it some weight so your blocker will stay in place. Stitch or glue the wide end closed, and voilà— you've made your house warmer and saved yourself a bundle of money. And in the summer, you can put your draft stopper to good use as a doorstop!

Draft Stopper, Take 2 🏠 ADD A LITTLE WHIMSY to your necktie draft stopper by turning it into a sea serpent. Just follow the directions for the Crafty Draft Stopper (above), then paint or sew a face on the wide end of the "body." If you want to go all out, cut triangles out of felt, and glue or sew them in a line along the back to form the monster's dorsal fins. Now your creature is ready to ward off sneaky winds.

Draft Stopper, Take 3 🏠 A BELOVED OLD JACKET that's way past its prime deserves a useful retirement. Use the jacket's sleeves in place of the tie in the instructions for the Crafty Draft Stopper (above). Cut them off below the shoulders, stuff them, and put them to work blocking drafts under doors.

That's Brilliant!

I hate air leaks in my home. They let in cold drafts in the winter and waste my precious money. Here are a couple of ways to tell if your home is leaking a little more air than it should be. If you can see light shining around the perimeter of your doors and windows, then they're letting the outside air in, and the inside air out. Or, close your windows and doors on a piece of paper. If you can pull the paper out without tearing it, then the seals aren't sealin' too well.

Crank Up the Shine
Brass and Copper Polish

> 1 tsp. of salt
> 1 cup of white vinegar
> 1 cup of flour

Beautiful brass and copper accents lend your home a lovely feeling of welcoming warmth. That soft metal sheen can add just the right amount of flair to an otherwise lackluster room. But if your metal is looking mighty dingy, it's not a pretty sight. So slather on this magic mixture to bring back the shine.

DIRECTIONS: Mix the salt into the vinegar until it dissolves, then stir in the flour until you have a smooth, thick paste. Sponge it onto the brass or copper, and let it sit—no elbow grease required! Wait about 15 to 30 minutes, depending on how tarnished the metal is, then simply rinse off the stuff with warm water, and buff with a soft, dry cloth to bring out that gleam.

Now, That's Hot! ☗ WANT TO GET that dull copper bright again, and whet your appetite at the same time? Just pull on a pair of rubber gloves, and use hot sauce as your polish! You can use whatever you have handy, whether it's a bottle of Uncle Jerry's Tongues of Flame (just kidding!) or that little packet of hot sauce left over from the taco joint. Pour the sauce on a soft, clean cloth, rub it over the surface, and let it sit for about 15 minutes before you wash it away with warm water.

Q *I bought a set of candleholders at a yard sale. The seller claimed they're brass, but I'm not so sure. How can I tell if they really are brass, or some other metal like bronze?*

A Here's how to tell: Brass, which is an alloy of copper and zinc, is a warmer, deeper yellow metal than bronze, which is usually a mixture of copper and tin. So if those decorative objects have a cooler hue, they're probably bronze. Inexpensive "brass" decorative objects, like pierced votive candleholders, are often made of bronze. Luckily, the solutions you use to spiff 'em up are the same for both metals.

Dazzling Basil Potpourri

When I found a box of clamp-lid canning jars in the attic, it brought back memories of this potpourri that Grandma Putt used to make with dried herbs and flowers from her garden. If you spent treasured childhood hours in your grandma's garden—or only wish you had—you'll love this mixer as much as I do. It has a simple, down-home aroma that no commercial air freshener can begin to match.

> 4 cups of sweet basil leaves and flower spikes
> 2 cups of dark opal basil leaves and flower spikes
> 2 cups of rosebuds
> 2 cups of rose geranium leaves
> 2 cups of rose petals
> 1 cup of lavender blossoms
> 1 oz. of powdered orrisroot*
> 1 oz. of sweet flag powder*

DIRECTIONS: Put the dried leaves and flowers in a big bowl, and toss them gently. Add the orrisroot and sweet flag powder, and toss the mixture again. Scoop the potpourri into clamp-lid canning jars or other airtight containers, and store them in a cool, dark place. To use your colorful, sweet-smelling creation, pour it into a basket or bowl, and set it out to be admired.

*Available in craft-supply stores, herb shops, and nurseries that specialize in herbs.

Florida Sunshine 🏠 FOR MY MONEY, nothing beats freshly squeezed orange juice. But what do you do with all those rinds? Here's what: Grate them, either by hand or in your food processor. Then dry the gratings on a screen, and add them to your favorite potpourri recipe. The citrus smell will make your house feel like the Sunshine State year-round.

PINCHING PENNIES

If your kitchen garbage can is starting to smell, you don't have to waste your money on those fancy air fresheners. Instead, try some potpourri for a scent-sational solution. To cover up unpleasant odors in your kitchen garbage can, fill a small, resealable plastic bag to within an inch or so of the top with your favorite potpourri. Then punch a few holes in the top of the bag, and tape the bag to the inside lid of the can.

End of Winter Window Cleaner

> **2 cups of rubbing alcohol**
> **1 tbsp. of ammonia**
> **1 tbsp. of dishwashing liquid**
> **2 qts. of water**

The outsides of windows pick up more than their fair share of grungy dirt over the winter. So as soon as the wild spring weather calms down, whip up a batch of this homemade miracle cleaner, and bring on the sunshine!

DIRECTIONS: Mix the ingredients in a bucket, then pour the solution into a handheld sprayer bottle, and go to town. This super-duper concoction will beat commercial streakless glass-cleaning products hands down!

The Simplest Cleaner 🏠 A LITTLE DISHWASHING LIQUID in a bucket of water works great to clean most windows, especially if you use a squeegee. Add only a small amount (about ⅓ tsp. per 3 gallons of water), so you don't get lots of suds or leave streaks behind when you wipe the cleaner off. You won't even need to rinse the glass unless it's absolutely filthy—just rub it dry with a soft, lint-free cloth and you're done.

Read All About It! 🏠 NEWSPAPER IS A TIME-TESTED window-cleanin' material. Just crumple a sheet of black-and-white newspaper—not the colored ads—and dip it into a shallow bowl of white vinegar. Rub the grime off your window with the newspaper, and finish up by drying and polishing the glass with a lint-free cloth.

GRANDMA'S OLD-TIME TIPS

Figuring out which side of the glass a streak is on can be awfully frustrating. You peer, you smear, you walk outside and back in again. To tell the difference at a glance, my Grandma Putt came up with a great trick: She used vertical strokes on the inside of her windows, and horizontal strokes on the outside. Then if that doggone streak was horizontal, why, she knew it was on the outside looking in!

Everything and the Kitchen Sink Cleaner

½ cup of borax
1 tsp. of ammonia
½ tsp. of dishwashing liquid
2 gal. of warm water

Looking for something you can whip up to tackle most of the major cleaning jobs in your house? Well, look no further. This simple formula is just the ticket for cleaning walls, floors, and just about every other washable surface in your house.

DIRECTIONS: Mix the ingredients in a bucket, then dip in your mop or sponge, and go to town! Whatever you're cleaning, it'll be spotless in no time at all.

Stone Cold Stone Floor Cleaning 🏠 KEEP A STONE FLOOR clean and sparkling by mopping it once a week with a solution made from a teaspoon or so of detergent-based dishwashing liquid per gallon of water. Rinse with clear water. (Don't use an all-natural soap-based product because soap will leave a residue that will make the stone look dull.)

Sponge on the Shine 🏠 TO REVIVE THE NATURAL SHINE on your stone floor, apply a little lemon oil (available at the hardware store) with a sponge or applicator mop. Wipe off the excess oil with an old towel, and buff your floor to a nice shine.

Get Down to the Nitty-Gritty 🏠 GRITTY DIRT can really do a number on a natural stone floor. You can strategically place floor mats so folks can wipe their shoes off before walking on the stone. But, let's face it—dirt happens. So use your trusty dust mop to fend it off, followed by a clean sweep with your vacuum cleaner.

That's Brilliant!

After you've installed new carpet in your home, stash the remnants in your workshop. Even the smallest pieces can come in mighty handy. You can glue tiny pieces of carpet, soft side down, to the bottoms of chair and table legs, so they'll glide easily, without leaving scratches or black marks on your stone and tile floors.

Fireplace Door Cleaner

½ cup of vinegar
1 tbsp. of ammonia
Water
Worcestershire sauce

Even with only occasional use, fireplaces can get pretty darn dirty. Clear away smoke and soot stains from your glass fireplace doors with a cloth dipped in white vinegar. But for more stubborn grime, try this mixture.

DIRECTIONS: Pour the vinegar and ammonia into a handheld sprayer bottle, and fill the balance with water. If the soot has baked onto the surface of the glass, buy a glass scraper at your local hardware store and carefully slice away the stains, then follow up with the vinegar and ammonia solution. Add a nice finishing touch by scrubbing the brass knobs with an old toothbrush—dipped in Worcestershire sauce, of course!

Scrape Off the Soot IF THE BUILDUP OF SOOT and creosote inside a fireplace gets too thick, it can actually catch fire. So to be safe, scrape the gunk off once or twice a season when the fireplace is cool. Remove the grate and the ashes, and line the floor of the fireplace with a plastic drop cloth topped with several layers of newspaper. Then scrape the sooty walls with a wire brush that's got a metal scraper attached to it, working from the top down. As debris collects on the floor, bundle it up in the top layer of newspaper and stuff it into a sturdy trash bag for disposal.

Super Shortcuts

Ashes can go flying everywhere when you sweep out your fireplace, so try my quick fix to make that chore less of a hassle. Simply sprinkle the ashes with damp coffee grounds. The flyaway ashes will clump up around the moist grounds, making it a cinch to sweep them up with your fireplace brush and dustpan. And play it safe—put the ashes into a metal container for extra insurance against an accidental fire.

First Response for Nail Polish Spills

1 part dishwashing liquid
1 part Kosher salt
2 parts water

There's no need to cry over spilled nail polish when you can use this easy trick to treat a fresh stain on your carpet or upholstery. First, blot up the excess polish so it doesn't get too sticky. Then whip up this concoction.

DIRECTIONS: Make a syrupy mixture by combining the ingredients in a small bowl. Pour the syrup directly onto the stain, and work it into a lather with a stiff brush. Blot and wipe the area, rinse the syrup off with a clean, damp cloth, and let the spot air-dry.

Make Mine a Double! ♠ IF A NAIL POLISH DISASTER hits your carpet, you need a double shot of alcohol. Saturate the stain with rubbing alcohol, and blot with a clean cloth to take the gunk near the surface off. Then soak the stain with cheap hair spray—which has a high alcohol content—and rub the deeper dye out. Keep folding the cloth so you don't spread the stain, and be sure to test the trick on an inconspicuous area first.

GRANDMA'S OLD-TIME TIPS

When nail polish landed on my Grandma Putt's painted walls, linoleum floor, or other hard surface, she harnessed the power of a 20-mule team. A single teaspoon of borax mixed in 1 cup of water did the trick, along with a big helping of elbow grease. Of course, Grandma always wiped up as much as she could while the spill was still fresh, before she called in the mighty mules. And, since I now have a granddaughter who's had more than her fair share of trouble with wayward nail polish, I know this remedy is still working as well as it did when Grandma used it.

Gold-Standard Sparkle Solution

All that glitters is not necessarily gold, but whether it's 24-karat or 10, when you have gold jewelry, you want your pieces to sparkle and shine like new. Here's a really simple potion to clean your precious gold. It takes just a minute to mix, and just one more minute to work its magic!

DIRECTIONS: Mix the ingredients in a small bowl that's deep enough to cover the gold pieces. Let them soak for 1 minute, then remove the items, rinse them under cold running water, and dry them with a soft, clean cloth.

24-Karat Smile 🔼 WHITE NON-GEL TOOTHPASTE is a gentle gold cleaner. Just squeeze a bit onto a soft cloth or an old toothbrush, and gently rub the pieces until they shine. Rinse thoroughly, using the toothbrush to coax paste out of the crevices. And check for loose stones before you begin because you don't want any baubles going down the drain.

Keep Away from Chlorine 🔼 CHLORINE DISCOLORS GOLD, eventually turning it black. So always wear rubber gloves when you're working with chlorine bleach, powdered cleansers, or other products that contain chlorine. And don't forget to take off your gold jewelry before you jump into the pool. And by all means, remove gold jewelry before you ease into a hot tub. The high water temperature in a hot tub is especially damaging to gold.

That's Brilliant!

You can scrub the dull film off gold by shaking it in a mixture of salt and vinegar or lemon juice. Start by filling a small bottle about three-quarters of the way with table salt, and then slowly pour in white vinegar or lemon juice almost to the top. Drop in your gold pieces, and set the bottle aside. Give it a shake several times a day, so the salt can gently scrub off any discoloration. After a week of shaking things up, empty the bottle, and thoroughly rinse your nice clean family jewels.

Handy Hearth Helper

Living in an area where winter seems to last from October until May really makes you appreciate how warm a fireplace can make your home. But the ashes and cinders that build up inside can get blown out onto the hearth, even if you use a screen. So try this recipe when you need to give a brick or stone hearth a good old-fashioned scrubbing.

4 cups of hot water

1 cup of soap powder (such as Ivory Snow®)

½ lb. of pumice powder*

½ cup of ammonia

DIRECTIONS: Put the water in a large bucket and mix in the soap powder. Slowly stir in the pumice and ammonia. Use a scrub brush—horsehair works best—to get deep into the grout between the bricks or stones. Rinse out the brush, and rinse the bricks by scrubbing them with clean water. Use a dry towel to soak up any excess water. Repeat the procedure until the bricks or stones are brand spankin' clean.

*Available at your local hardware store.

Home Is Where the Hearth Is

IF YOU'RE LUCKY ENOUGH to have a fireplace, it's time to learn how to clean it! For brick fireplaces, combine 1 ounce of dishwashing liquid and 1 ounce of salt with a little water. The mixture should be fairly thick. With a soft sponge, apply it to the brick. Let the solution stand for 30 minutes, then remove it with a sponge soaked in warm water. If you have a stone fireplace, add 2 tablespoons of dishwashing liquid to a bucket of water. Sponge the mixture onto the fireplace, working from the top down. Dry with a soft cloth.

PINCHING PENNIES

I used to love my old red Radio Flyer® wagon when I was a kid, and they're still pretty popular today. But when your children outgrow the old wagon, don't roll it to the curb. Instead, use it to lug fireplace wood in from the great outdoors. It'll save you a few trips out into the cold and help you keep your house toasty warm.

Herb Garden Potpourri

Why spend a bundle for fancy, store-bought air fresheners when you can make your own from ingredients right in your garden? For a potpourri that smells good enough to eat, give this magical mixture a try.

> 2 cups of thyme shoots
> 1 cup of dried mint leaves
> 1 cup of dried rosemary leaves
> ¼ cup of whole cloves
> ¼ cup of dried lavender flowers
> 2 tbsp. of powdered orrisroot*

DIRECTIONS: Mix the ingredients in a large bowl with your hands (they'll smell great afterward). Sew the mixed potpourri into fabric sachets and use them to freshen up closets and drawers. Scoop the leftovers in closed glass jars for storage.

*Available in craft-supply stores, herb shops, and nurseries that specialize in herbs.

A Place for Everything

WHETHER YOU USE HERBS for freshening the air, cooking, or making healing salves, the worst place to store them is in the kitchen on an open shelf, or anywhere near the stove. That's because heat and light dry up the volatile oils, making the fragrance and flavor go downhill fast. The ideal storage space is in a cool, closed cupboard. But if you're short on cupboard space, just tack a strip of elastic to the inside of one of your kitchen drawers. Then tuck the herb bottles under the stretchy band. They'll be easy to reach—and will keep their flavor a lot longer!

GRANDMA'S OLD-TIME TIPS

My Grandma Putt had an old trick to sweeten the air in her house: She'd dab a little perfumed oil on a cotton ball, and stuff it into the heating vent. When the heat came up, it would scent the whole room. She also liked to set little bowls of vanilla extract around the room, though some of you might find that to be an awful waste of a pricey baking ingredient. If so, then stick with setting out bowls of aromatic potpourri, such as the Herb Garden Potpourri (above).

High-Powered Dirty Door Handle Helper

> 3 parts ammonia
> 1 part vinegar

Grubby hands are touching them all day long, so doorknobs and cabinet handles build up more than their fair share of dirt and germs. A routine swipe with plain soap and water will do the trick on most knobs and handles, but if yours are really grimy, call on the magic power of this mixture to make that crud vanish in a jiffy.

DIRECTIONS: Combine the ammonia and vinegar in a handheld sprayer bottle. Then spray some of the mixture on a damp microfiber cleaning cloth, and give your cabinet handles and doorknobs a thorough wipe-down, working on them one at a time. Remember to reach inside of any hollowed-out handles (where greasy residue can hide), and rinse off each handle with a clean, damp cloth before you move on to the next one. And don't forget to keep a clean, dry cloth or paper towel within easy reach—that way, you can catch any runaway drips as they start, so they don't leave streaks on your doors!

One Cleaner Fits All 🏠 SOAP MIXED WITH WATER is a good all-purpose cleaner for nearly every kind of exterior door, except for unfinished wood. Just wet a sponge in a bucket of soapy water, squeeze it out, and wipe the door with up-and-down strokes. Rinse the door with clean water, and wipe it dry. This method works well on finished wooden doors because the quick treatment doesn't allow moisture to soak in.

PINCHING PENNIES

During the height of cold and flu season, when everyone seems to have the sniffles, it's a good idea to wipe down your doorknobs every day to kill the germs. But don't spend money on expensive store-bought sanitizer. Instead, reach for plain old rubbing alcohol. It does such a great job of getting rid of nasty germs that surgeons use it to sterilize their hands. Just pour a bit on a clean cloth, and wipe the knob thoroughly. (Be careful not to get the alcohol on any wood door.) And there's no need to rinse because the alcohol will evaporate lickety-split, leaving a germ-free surface behind.

Let There Be Light Lamp Shade Shampoo

¼ cup of dishwashing liquid
1 cup of warm water

One of the easiest ways to add a little pizzazz to a dull room is to put pretty new lamp shades on your old lamps. Even cleaning your existing shades will brighten up the room, since all that dust and dirt blocks the light. It's easy! But lamp shades made of paper or parchment are usually glued together, so don't even think about dunking them under water to clean them. Instead, give them this special suds-only shampoo.

DIRECTIONS: Use an electric mixer to whip the ingredients into a thick foam (the consistency of whipped cream). Sponge just the suds over the inside and outside of the shade, and rub gently. Wipe the soap off with a clean, damp cloth and dry the shade with a soft, clean cloth. This technique works great on fabric shades, too, and on shades that have glued-on fancy trim. Just keep in mind that lamp shades must always be handled very carefully; otherwise, you'll make dents or creases in them.

Hit the Spot 🏠 THE FAST AND EASY WAY to remove dirty fingerprints from lamp shades is by rubbing them gently with a baby wipe. It takes just a few quick strokes to get rid of most grungy smears. Then make sure you wipe down the rest of the shade after you make the fingerprints vanish, so the clean spots don't stick out like a sore thumb.

That's Brilliant!

You can use the upholstery brush attachment on your vacuum cleaner to suck the dust off lamp shades. Or take the opposite tack, and try a blast of compressed air to blow the dust out of those pleats and folds. And while you're at it, use that high-pressure air to get the dust out of the nooks and crannies in fancy lamp bases, too.

Magic Marble-Brightening Poultice

> 3 or 4 tbsp. of whiting
>
> 2 tbsp. of banana oil

The "whiting" in this recipe is calcium carbonate that's been ground to a fine powder. What's calcium carbonate, you ask? It's nothing more than plain old white chalk. But when it's mixed with a stain-removing chemical to make a paste, it's called a marble poultice, which can make stains vanish without harming the marble. You'll find the banana oil (amyl acetate) at the drugstore; it's not made from bananas, but it smells just like them!

DIRECTIONS: Stir the whiting into the banana oil, adding 1 tablespoon at a time until it becomes a creamy paste. Spread the poultice onto the stain, and cover it with plastic wrap. Use masking tape to hold the plastic wrap in place. Remove the poultice after about 30 minutes, and wipe off the residue with a moist cloth. **Note:** Be careful when you work with the banana oil because it's highly flammable—don't use it near a flame or spark. Also, avoid breathing the fumes, no matter how good you think they smell, and keep a window open to ventilate the area.

Acid Is the Enemy

MARBLE IS A POROUS STONE, which means it can absorb spills and stains quite easily. But what's even worse is that marble can actually be dissolved by acid. That's why it's important to wipe up spills as soon as they happen. Why, even something as seemingly harmless as fruit juice or cola can contain enough acid to eat into your marble. So keep cloth or paper towels handy, and always use coasters on marble surfaces.

Super Shortcuts

To remove mild stains from marble surfaces, sprinkle some baking soda onto a damp cloth, and rub the spots away. For more stubborn stains, make a paste of baking soda and water, apply it to the stains, and let it sit for about an hour before rinsing it away. And don't worry—baking soda is alkaline, not acidic, so it won't eat into the marble.

No More Streaks Glass Cleaner

> 1/3 cup of white vinegar
>
> 1/4 cup of rubbing alcohol
>
> 3 cups of water

Do you ever wonder why some cleaning professionals have the reputation for saying "I don't do windows"? Maybe it's because of the smell of the ammonia that's in most powerful window-cleaning formulas. This easy streak-free solution will make your windows and other glass surfaces sparkle without the in-your-face ammonia aroma.

DIRECTIONS: Mix the ingredients in a handheld sprayer bottle, and shake well. Spray the solution onto a soft microfiber cloth, or directly on the smudgy glass surface, and swipe it clean. Your windows will be the cleanest they've been since they were brand new.

It May Sound Corny, But . . . ⌂ IF THERE'S A SMOKER in the house or you prepare a lot of fried foods in the kitchen, cleaning your windows probably requires a little extra effort. Mix some mildly abrasive cornstarch with ammonia and water to clean grime-clouded windows. Then rinse it all off with clear water.

Give Your Windowsills a Rubdown ⌂ WINDOWSILLS CAN GET really grimy, and sometimes using detergent and water isn't enough to keep them clean. So try cleaning the sill with a cloth dipped in rubbing alcohol first. It should dissolve the dirt enough for the detergent to get down and do its thing. **Note:** Use only on painted windowsills; alcohol can take the finish off stained wood.

GRANDMA'S OLD-TIME TIPS

Grandma Putt always washed her windows from the top down, to avoid leaving them with water spots. Every once in a while, though, she did wind up with water spots here and there. When that happened, she wiped kerosene onto the marks with a soft cotton cloth, then rubbed them with crumpled newspaper. Bingo! No more spots!

Old-Time Board Cleaner

You've found some beautiful old wood boards that you want to recycle into paneling or furniture. There's just one problem: The things are filthy! Well, don't fret about it. Just mix up a batch of this power-packed cleaner. It—along with a little elbow grease—will make that vintage lumber look better than new.

DIRECTIONS: Mix the ingredients together in a bucket, and scour the boards using a stiff scrub brush. Rinse with clear water, and rub dry with a clean towel. Then pat yourself on the back for your lucky find because new wood could never look this good!

Painted-Wood Cleaner 🏠 GRANDMA CLEANED PAINTED-WOOD surfaces like floors, furniture, and woodwork with a solution of 1 teaspoon of baking soda per gallon of hot water. She applied it with either a mop or a sponge, then wiped the surface dry with a soft cloth. (To dry a wooden floor, she wrapped the cloth around the business end of a dust mop.)

That's Brilliant!

Yikes! You spilled paint—and not only that, but the paint is oil-based, the "victim" is your favorite wooden table, and you didn't reach the scene until the spots had dried. Well, stiffen up that upper lip, and reach for a bottle of boiled linseed oil. Brush a generous coat onto the marks, and let it stand until the paint has softened (how long this takes will depend on how long the paint has been there). Then remove it with a soft cloth soaked in more boiled linseed oil. Finally, scrape off any residue with a plastic scraper or an old credit card. Whatever you do, don't use paint remover or thinner, or you'll ruin the wood's finish.

One Gel of an Air Freshener

Just about everybody has at least one box of drinking glasses stashed in the attic. Whether yours are unused wedding presents, jelly glasses with Howdy Doody prancing across the sides, or souvenirs of your 1966 road trip, you can turn them into clever gifts with the help of this simple recipe.

> **2 cups of distilled water**
> **4 packages of unflavored gelatin**
> **10 to 20 drops of fragrant oil**
> **Food coloring**

DIRECTIONS: Heat 1 cup of the water almost to boiling, then add the gelatin and stir until it's dissolved. Remove from the heat and add the remaining cup of water, the fragrant oil, and the food coloring (enough to get the shade you want). Pour the mixture into clean glasses, and let them sit at room temperature until the gel is fully set. **Note:** Although the gel will set faster in the refrigerator, it'll also share its scent with your food, so it's best to have patience and let the mixture cool in the glasses at room temperature.

Hand Wash Your Good Glass

DISHWASHERS ARE QUICK and easy, but they can etch and even dull your lead crystal or gold- and silver-rimmed glassware. So when you need to wash the good stuff, do it by hand—unless you're a real butterfingers. Start by padding the sink with a rubber mat, and lay a bath towel over the front edge. Move the faucet out of the way, too, so you don't accidentally whack the Waterford. Then wash each piece in warm, soapy water and rinse with the sink sprayer.

PINCHING PENNIES

Want your glassware to really sparkle without spending too much on cleaners? Then add white vinegar to the rinse! If you're washing by hand, add a capful to a sink of clean water; for dishwashers, pour ¼ cup into the rinse cycle. White vinegar counteracts the film that hard water can leave behind, and your glassware will be squeaky clean with no streaks or spots.

Paneling Perker-Upper

Commercial oil soaps do a fine and dandy job of heavy-duty cleaning, but you can make your own at home for just a fraction of the cost. Use this potion to clean greasy, grimy buildup off your wood or faux wood paneling, kitchen cabinets, and wood furniture. It'll cut through the oily dirt in a flash and leave behind a fresh-scented shine.

> ¼ cup of lemon-scented dishwashing liquid
> 1 tbsp. of olive oil
> 1 qt. of warm water

DIRECTIONS: Mix the ingredients in a bucket, moisten a clean, dry cloth with the solution, and rub your paneling, starting at the bottom and working your way up. There's no need to rinse; just buff the paneling with a clean, dry terry-cloth towel to bring out the shine. Work in sections, but don't leave any of the solution on the wall for longer than a minute or two, or it could penetrate the finish and leave a cloudy haze. And buff the wood with the grain, not in circles, so your stroke marks blend in with the wood.

Waterproofing Wood 🏠 WHEN IT COMES TO bringing a special glow to your dining room table, or anyplace else in the house for that matter, nothing beats good old-fashioned candles. But these waxy wonders can do a whole lot more than just light up your life—even when they've burned down to mere stubs. You can use 'em to keep the cut edges of plywood and other lumber from absorbing moisture (including the tops of your interior doors). Just rub the edges with a candle and they'll be waterproof.

Super Shortcuts

✂ Before you buy any type of wood furniture, you need to determine what kind of wood it's made of. There are various grades and types of wood on the market, ranging from bargain-basement particleboard to exotic (and expensive) hardwoods. Once you know what type of wood you want, shop around on the Web to determine what a fair and reasonable price is. You'll save yourself a lot of time and trouble by planning ahead.

Perfect Paneling Potion

> 1 cup of white vinegar
> ½ cup of olive oil
> 1 gal. of warm water

There's no escaping dirt on your walls, no matter what they're made of. Even if your walls are covered with paneling, dirt will find a way to accumulate. So brighten drab paneling by dusting first, then washing the walls with this grime-grabbing formula.

DIRECTIONS: Mix the ingredients in a bucket. Use a large soft cloth to apply the formula to the paneling. Allow it to air-dry for a few minutes, then buff the surface with a clean, dry cloth.

Just the Stuff for Scuffs 🏠 IF THE BACK of your chair leaves a scuff on the wall, here are a couple of tricks you can try to remove the mark: Use an old, clean toothbrush to scrub the mark with white non-gel toothpaste. Or, erase it with a rubber eraser. Use either the soft white rubber or the crumbly gum kind to erase most scuffs without damaging the paint or wallpaper. But don't try this with a pink or any other color eraser, because it may leave a new stain on the wall.

Up Against the Wall 🏠 IT'S EASY ENOUGH to wash painted walls, but what about wallpapered walls? After all, they accumulate dirt and dust just like any other surface. But some wallpaper can't be washed at all. To determine if yours is safe to clean, select an inconspicuous area to test. Wash the spot with a little baking soda and water. If it's safe, proceed from the top down, just as you would with painted walls. If the paper starts to discolor, abandon the wet wash and simply dust with a soft, dry cloth.

PINCHING PENNIES

If you're like me, you've got old wallpaper odds and ends just sitting on a shelf somewhere. Wallpaper isn't cheap, so you can't bear to throw it out, right? But what the heck are you ever going to do with the stuff? Here's one solution: Make placemats by cutting the wallpaper to the desired size, covering the front and back with clear Con-Tact® paper, and trimming, leaving a ¼-inch margin of Con-Tact paper on all sides. This works great, and looks great, too—especially if the mats match the paper on your dining room walls!

Plaster-Texturizing Tonic

2 to 3 cups of fine white sand
1 gal. of interior latex paint

Sand paint has one feature that really makes it a worthwhile choice for ceilings: its texture. Use this gritty paint to hide small cracks and other surface flaws. You'll wind up with an attractive pattern, too. So, if your ceiling is blah and bland, add a little excitement with this concoction.

DIRECTIONS: Pour the sand through a piece of wire mesh (like a window screen) to remove any grains that are too coarse to use. Pour the paint into a 2-gallon bucket, stir in the sand, and mix the concoction together thoroughly. Just make sure to take note of how much sand you use so you get the same mixture for each successive gallon of paint.

Keep the Lid Clean 🏠 GETTING READY TO PAINT?

Before you start, tape a piece of wire (any kind will do) across the top of the opened paint can at the center. Then wipe your paint-loaded brush against the wire, rather than the side of the can. This way, paint won't collect in the rim and make the lid stick when you close the can.

Paint Around Fixtures 🏠 IF YOU'RE GOING to be

painting a wall, prepare the area first so you don't get paint in places it's not supposed to be. Spread drop cloths over furniture, even if you've pushed it into the middle of the room. Use painter's tape and plastic sheeting or garbage bags to protect fixtures, light switches, and woodwork. Then get to work and paint!

✂ Super Shortcuts

When you want to hand-paint small objects, save yourself a lot of time and wasted effort by attaching double-faced tape to a large piece of cardboard, and stick the little things in place on the tape. They'll stay put as you work—and your fingers will stay clean.

Put Spice in Your Life Potpourri

Grandma made sure the smell of fresh potpourri always permeated her house. It never failed to calm her down—even when I managed to get her good and riled up. To make her favorite potpourri, Grandma Putt started with a lidded, quart-size glass jar and a handful of aromatic ingredients.

4 lightly packed cups of fresh rose petals

3 tbsp. of mixed spices like ground cloves and allspice

1 tbsp. of powdered orrisroot*

DIRECTIONS: Place all the ingredients into the jar and mix well. To use, set the potpourri out in decorative dishes, and in just a few minutes you'll have a wonderfully pleasant aroma going through the entire house. Store any remaining potpourri in the jar.

*Available in craft-supply stores, herb shops, and nurseries that specialize in herbs.

Protect Your Fingers 🏠 WHEN YOU'RE CUTTING ROSES to take inside, protect your fingers the way Grandma Putt did. Just squeeze the thorny stem with a clip-type clothespin, and you won't have to worry about the little stabbers.

So Long, Smoke! 🏠 WHEN YOU NEED TO REMOVE the smoky smell from a room, fill a glass jar with vinegar (any kind will do), and sink a large oil-lamp wick into it. The wick will absorb the noxious odors, and the vinegar will knock 'em dead. By the next day, you'll be able to enter the room and forget all about yesterday's smoky scene.

PINCHING PENNIES

What do you do with a fresh Christmas tree after the holidays are over? One of my favorite reuses is to put the tree to work freshening the air. Here's how: Pile small cuttings in a basket and place it on the floor of your bathroom to keep the air from turning foul. You can also use dried sprigs as kindling for a cozy fire; they give off that terrific piney scent. Or, place evergreen needles in dishes and set them throughout your house. Whenever the air could use a perk-up, just stir the needles to release their aroma.

Shine Like the Moon Jewelry Bath

Who doesn't love it when their jewelry shines so brightly that folks can see the sparkle from clear across the room? Give 'em something to talk about with this good-as-gold solution.

DIRECTIONS: Mix the ingredients in a glass bowl, and soak your pieces in it for about five minutes. Use a very soft brush to clean the crevices, then rinse the jewelry off in clean water, and dry it with a soft cloth. **Note:** Don't use this tonic if your pieces include any soft stones, like pearls or lapis, because they might be damaged by the ammonia.

Finishing Touches 🏠 MAKEUP, HAIR SPRAY, and other lotions and potions will dull your jewelry, so always wrap up your beauty routine before putting on your fancy doodads. And to retain your jewelry's razzle-dazzle, make sure you wipe it with a soft, clean cloth after you take it off to remove any traces of perspiration, perfume, or other dulling substances.

GRANDMA'S OLD-TIME TIPS

*When your silver jewelry starts looking a little dull, bring back its good-as-new glow with this little trick from Grandma Putt. Sprinkle baking soda over the bottom of an aluminum foil baking pan, place your jewelry in the pan, and then pour enough boiling water over it to completely cover the pieces. Let them soak for two to three minutes, and the tarnish will simply disappear. Remove your jewelry using plastic tongs (so you don't burn your fingers or scratch the metal), rinse it under cold water, and buff it dry with a soft, clean cloth. **Note:** Boiling water is safe for silver, but not for any kind of stones (real or fake), so use this trick on all-silver pieces only.*

Silver Jewelry Sparkle Solution

> 2 tbsp. of borax
> 2 tbsp. of dishwashing liquid
> 1 tbsp. of ammonia
> 2 cups of hot water

Restore the sparkle and shine to your sterling silver jewelry with this DIY dip. You can whip it up in a jar so you'll have plenty on hand whenever you need it. That makes it as convenient as any store-bought cleaner, but it costs only pennies to make.

DIRECTIONS: Mix all of the ingredients in a bowl, stirring gently so foam doesn't form. Gently pour the solution into a glass jar with a tight-fitting lid. To clean sterling silver jewelry, soak the pieces in the solution for 5 to 10 minutes, and brush any residue away with an old, clean toothbrush. Rinse the pieces under warm water and dry them with a soft cloth. **Note:** Do not clean pearls or emeralds with this solution because they might be damaged by the ammonia.

Keep Rubber Away 🏠 RUBBER CAUSES A CHEMICAL reaction that tarnishes silver and can even corrode it, so keep it away from this precious metal. Don't use rubber gloves when you wash or polish silver, and never put a rubber band around flatware. Even a rubber seal inside of a storage chest or a stray sealing ring from a canning jar may affect the finish.

Super Shortcuts

✂ Simple and powerful—now that's the kind of magic I like! And what could be a simpler silver-cleaning solution than plain old baking soda? Just make a paste of baking soda and water, dab some onto a soft cloth, and rub the tarnish off your silver. Rinse the piece thoroughly, and rub it dry with a soft, clean cloth. Keep in mind that baking soda is a very mild abrasive, so if you're worried at all about scratching the family heirlooms, ditch this old-time cleanser and use a commercial silver polish instead.

Sink the Stink Room Spray

Let's face it: Some rooms are stinkier than others, so keep a bottle of this air freshener wherever you need it most. But why wait for the stink? You'll probably find yourself spraying this nontoxic air freshener just because it smells so good.

1 cup of distilled water
1 cup of rubbing alcohol
2 drops of orange oil

DIRECTIONS: Mix the ingredients in a handheld sprayer bottle, and give the room a quick spray or two whenever you need to clear the air.

Mouse-Cage Freshener 🏠 WHEN I WAS A BOY, I kept a few white mice as pets. They were great little pals, but there was one problem: No matter how clean I kept their cage, the little rodents always gave off a distinctive odor. To absorb the aroma, Grandma had me put a bowl of vinegar next to the cage (but not in it). I replaced it with a fresh supply every few days, and my room always smelled fresh!

Breathe Easy 🏠 THE EASIEST WAY to clean the air in your house is by replacing the filters in your furnace or air conditioner. And since disposable filters only cost a few bucks, buying replacements shouldn't put too big a dent in your wallet. Still, if you're of a rather thrifty nature and want to save a little money (and who doesn't?), buy your disposable filters by the case, instead of one at a time.

That's Brilliant!

You already know how well vinegar works as an air freshener in your home. Well, it works just as well in the car. To rid your ride of any unpleasant odor, pour a little vinegar into a shallow bowl and set it in your car overnight. Just remember to take it out before you start driving away, or you'll have vinegar sloshing around your car's interior!

So Long, Soot Fireplace Scrub

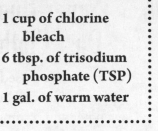

Many people who live in a cold climate love to keep a roarin' fire going all winter long. Come spring, the inside of your fireplace will be so thick with built-up soot that you'll need to mix up a batch of this super soot-scrubbin' solution.

> 1 cup of chlorine bleach
>
> 6 tbsp. of trisodium phosphate (TSP)
>
> 1 gal. of warm water

DIRECTIONS: Mix the ingredients in a bucket. Line the floor of the fireplace with a drop cloth and newspapers, and set the bucket inside. Dip a wire brush into the solution and scrub the blackened walls, working from the top down. Wear rubber gloves, safety glasses, and protective clothing to guard your skin from accidental spills and splashes, and be sure to use plenty of the solution—along with a big ol' helping of elbow grease!

Finesse Fireplace Furnishings

CLEANING YOUR FIREPLACE GRATE and wrought-iron tools is messy, so haul them outdoors to do the job. Scrub the soot buildup off the grate and the andirons with a wire brush. Then rub the grate, andirons, and fireplace tools with a fine (000 grade) steel wool pad dipped in mineral oil. Wipe the excess oil off with a dry cloth, and your furnishings are ready for action.

Good-to-Go Gas Logs

TO CLEAN GAS FIREPLACE LOGS, make sure they're cool, then dust them monthly with a soft brush. Be extra careful when you're cleaning around the gas vents so you don't nudge them out of position and cause them to malfunction.

Super Shortcuts

✂ Some soot stains can be impossible to remove completely, no matter how hard you scrub. But don't give up—try freshening up the fireplace with a coat of paint instead. You'll find plenty of fireplace-painting kits at your local hardware store. But before you go gung ho, remember that nothing can really match the warm look of bare bricks—and painting is pretty permanent.

Super-Duper Souped-Up Dust Cloths

Grandma Putt used these powerful picker-uppers to dust all her wooden furniture, and even her stair railings. I still use 'em to this day. Besides making the wood gleam, they help condition it with every wipe.

> 2 tbsp. of boiled linseed oil
> 1 tbsp. of ammonia
> 1 tbsp. of soap powder
>   (such as Ivory Snow®)
> 1 qt. of warm water
> 1 soft cotton cloth*

DIRECTIONS: Mix the linseed oil, ammonia, soap, and water in a small bucket, and soak the cloth in the mixture for four or five minutes. Wring out the cloth, hang it up to dry, and store it in a glass jar or plastic container with a tight-fitting lid. After you've used it for dusting, wash it as you would any rag, and re-treat it with the formula.

*A cloth diaper or a piece of an all-cotton flannel sheet is perfect.

Dust Cloth Sock 🏠 HAVE YOU EVER WONDERED what becomes of all the single socks that go astray between the laundry hamper and the end of the drying cycle? Well, I have, and even though I've never solved that mystery, I do know a great use for the lonely left-behinds. A cotton sock, pulled over your hand, is just the ticket for dusting knickknacks, mini or venetian blinds, or anything else for that matter! To extend your reach, or to get into tight places (say, under the couch or clothes dryer), slip the sock over a yardstick or broom handle.

PINCHING PENNIES

If you think of a chamois as just another part of your car wash kit, think again. These soft leather marvels are also perfect dust cloths that will not scratch delicate surfaces, like telescope and camera lenses, or even photographs. The most delicate of items can still be kept clean with a chamois or two close at hand.

Super Soot Remover

> 1 bar of Fels Naptha® soap
> 3 qts. of water
> 1 cup of ammonia
> 1 pound of pumice powder

Make black stains disappear from the brick around your fireplace with this super scrub—it gets its power from pumice, a powdered stone you can buy at your local hardware store. Before you start, cover the floors and furniture near the fireplace with plastic drop cloths to protect them from drips and splashes. And keep in mind that soot stains are notoriously hard to remove, so even after using this powerful potion, some blackened areas may still remain.

DIRECTIONS: Slice the soap into the water in a deep stockpot, and bring it to a boil. Reduce the heat and simmer the mixture until the soap melts, then remove it from the heat and let it cool. Stir in the ammonia and the pumice, mixing thoroughly. Brush the glop onto the stained brick, let it sit for about an hour, and then scrub the area with a stiff brush. Rinse the mixture off, and let the clean brick air-dry.

Stinky Fireplace? SOUR, SMOKY SMELLS coming from your fireplace are a sure sign that moisture has seeped into the chimney and soaked into the sooty deposits inside. The first step is to clean your chimney to reduce that crud—it's not only smelly, but also can be dangerous! For a permanent solution, have a chimney cap with a top damper installed to block the rain and snow, which will also keep bad smells from sinking down your chimney when the air is cold.

Q *I need to buy cut wood for my fireplace, and I'm comparing prices from several vendors. They all sell their wood "by the cord." Just what is a cord, anyway?*

A A cord is simply a measure of volume. It's 128 cubic feet to be exact. More simply put, a cord of wood is a rectangular stack of wood that is 4 feet wide, 4 feet high, and 8 feet long. Another common term is *face cord*, which is half a cord—2 feet wide, 4 feet high, and 8 feet long. Now you know!

Terrific Toy De-Grimer

With little ones in your home, it's hard enough just to keep things picked up. Who has time to clean their kids' toys? Don't ignore it for too long—it doesn't take much for grime and germs to pass from one toy (or kid) to another. So scrub hard plastic or soft vinyl toys clean with this simple mixture. It works great on vinyl dolls, squeeze toys, rattles, and toy cars.

> ¾ cup of baking soda
> 1 tbsp. of dishwashing liquid

DIRECTIONS: Mix the ingredients in a small bowl to make a paste. Scoop some onto an old toothbrush, and scrub the dirt away. Rinse thoroughly under running water while wiping with a sponge or soft cloth to remove any residue.

Easy "Painting" 🏠 LIKE ANY LITTLE KID, I always wanted to help with whatever work Grandma and Grandpa were doing around their house. They encouraged my enthusiasm—even when it came time to paint the front porch. Grandma just gave me a clean paintbrush and a small pail filled with water, and let me "paint" the steps. Because the water made the wood look darker, I thought I was really accomplishing something. By the time the stairs had dried to their normal color, Grandma had put me to work "painting" the fence.

That's Brilliant!

Here's a quick and easy way to clean a baby rattle, a rubber ducky, and any other grubby plastic toy—just run them through the dishwasher. Place rattles in the silverware basket, set larger toys on the top rack, and fasten small ones to the rack with twist ties, so they don't get dislodged during the cycle. Press the button, and soon the toys will be squeaky clean. Not all toys can handle very hot water, so if you're concerned that a toy might melt, wash it by hand instead.

Ultra-Simple Window-Washing Solution

Have you noticed that sunlight just isn't pouring into your house the way it used to? It could be that your windows are getting a little too grimy, and it's time for a good cleaning. Nothing is easier to mix than this solution that'll clean every window in your house for pennies—but it'll look like you hired a window-washing pro.

DIRECTIONS: Pour the vinegar into a handheld sprayer bottle, then fill the jar the rest of the way with water. Spray it on your windows, inside and out, and wipe them dry with a lint-free cloth or crumpled-up newspapers.

Power Wash ⬆ EXTERIOR WINDOWS get a lot grubbier than their inside cousins, so rinse off as much dirt as you can with a strong spray from a garden hose before you begin scrubbing. This will wash off surface grime, knock down cobwebs, and help loosen the dirt before you apply the elbow grease. If the windows are really filthy, rub the worst of the dirt off with a nonabrasive plastic scrubbie. Then follow up with a sponge or cloth dipped in your favorite window cleaner, and squeegee them to perfection.

Window Wash ⬆ MAKE A TERRIFIC streak-free window cleaner by mixing equal parts of rubbing alcohol and sudsless ammonia in a handheld sprayer bottle.

PINCHING PENNIES

In addition to being just about the best window-cleaning tool out there, here's another great reason to invest in a good-quality squeegee: Most of them are made to screw onto an extension pole, allowing you to reach the tops of tall windows without a ladder. So if you have high windows, an extension handle is well worth the investment. But before you buy one, check the broom closet—you may already have a long-enough broom or mop handle that can be unscrewed and used with the squeegee.

Wake Up and Smell the Windows Spray

Bees and butterflies love the long-lasting, spiky purple flowers of anise hyssop (*Agastache foeniculum*)—and so do I! But this perennial is much more than just a pretty face; its fragrant leaves make a great-smelling addition to my home-made window cleaner.

DIRECTIONS: Place the anise hyssop in a quart-size glass jar, then pour in the vinegar. Put the lid on the jar, shake, and let the mixture sit for a week. (During that time, shake the jar daily, and add more vinegar, if needed, to keep the leaves and/or flowers covered.) Strain, then add ⅓ cup of the remaining liquid to the water, and pour into a handheld sprayer bottle. Spray the windows, and wipe them clean with a soft, dry cloth.

Go Lint-Free 🏠 RUBBING WINDOWS with a wad of paper towels will leave behind a telltale trail of lint, so use a soft, lint-free cloth instead. Old cloth diapers are a great choice because they're highly absorbent and super soft. Or you can use a microfiber cloth or chamois. And skip the fabric softener when you wash window-cleaning cloths: It reduces their absorbency and may leave streaks on the windows.

Super Shortcuts

✂ A good squeegee makes short work of window cleaning, so ignore the cheap plastic ones you can pick up at discount stores. Spend a little more money and invest in a professional squeegee for about $10—it's well worth the price. Because the soft rubber blade has no imperfections, it'll do a streak-free job, and you can even replace the rubber strip when it wears out. A 12- or 14-inch blade is a good size for most home windows. You'll find professional-quality squeegees at all janitorial-supply companies and most hardware stores.

Wall-Repairing Plaster Paste

3 tbsp. of cornstarch
3 tbsp. of salt
2 tbsp. of water

Uh-oh! The kids got a little too rambunctious in the family room, and now there's a dent in the middle of the wall. Take a deep breath, calm down, and head to the kitchen. There's no need to purchase a special product to fill that hole. Just whip up this paste—you'll never notice the difference.

DIRECTIONS: Mix the cornstarch and salt together, then add the water to make a thick paste. Using your fingers, smear it on your walls wherever you have a nail hole, chip, or dent. Allow the paste to dry, then use a damp sponge to remove any excess paste and make the surface smooth and even. Once the area is totally dry, paint it to match the wall.

Handling Heavy-Duty Dirt 🏠 IF PAINTED WALLS have been neglected for years, or if they're stained with nicotine or soot, trisodium phosphate (TSP) will get rid of the accumulated dirt and discoloration. TSP is strong stuff, so test it in an inconspicuous area first to make sure it doesn't damage the paint. Then mix 1 to 2 tablespoons of TSP powder in a gallon of warm water, wash the walls from the bottom up, and rinse them thoroughly. Repeat the treatment until you're satisfied that the walls are clean. And be sure to wear rubber gloves and protective clothing, and cover all rugs, floors, and furniture to shield them from drips and splashes.

That's Brilliant!

Wall washing goes a lot faster when you use two buckets: one for the cleaning solution and the other for clean water to rinse the sponge out. Work from the bottom up, and rinse each section before going on to the next. Change the water frequently, so you aren't wiping dirt back onto the walls as you go. And check for streaks when you finish; if you see any, give them another going-over to blend them in.

Water-Mark Wonder Wipe

> **1 part white vinegar**
> **1 part olive oil**

If you're lucky enough to have a beautiful wooden table, I'm sure that you take every precaution to keep it looking its best. But not everyone is as careful as you are, so what if someone decides to set a wet glass on the table without a coaster? Here's a no-sweat solution that'll erase water marks with a single swipe.

DIRECTIONS: Mix the ingredients together and rub them into the telltale white rings with a soft cloth, using a circular motion. Then buff the table with a second cloth. Those water marks will be gone in no time.

Wooden Table Care 🏠 YOU CAN REMOVE WATER MARKS from wooden tables with this easy method: Put 4 tablespoons of virgin olive oil and 3 tablespoons of paraffin shavings in a double boiler, and heat until the wax has melted. Remove the pot from the heat, stir to mix the ingredients, and let cool. Then dip a clean, soft cotton cloth in the paste, and rub it into the spotted area, using a circular motion. Buff the area to a shine with a second cloth.

Car Wax Cleanup 🏠 HERE'S A SIMPLE METHOD for removing white water-mark rings from wooden furniture: Just grab a can of car wax from the garage, dip a soft cloth into the wax, and gently rub the marks away.

GRANDMA'S OLD-TIME TIPS

Grandma had a simple method for getting white water rings off wooden tabletops. She used a pinch of salt and a drop of water, rubbing it into the wood with a soft cloth until the stain disappeared. Then she followed up with her regular furniture polish. Her tables were so spotless after the salty rub, you could see your own reflection in them!

Wonder-Working Wall Wash

When you're cleaning house, don't forget about your walls. This terrific tonic will make the greasy film, dirty fingerprints, and general grubbiness simply vanish like magic.

> 1 cup of ammonia
> ½ cup of white vinegar
> ¼ cup of baking soda
> 1 gal. of warm water

DIRECTIONS: Mix the ingredients in a bucket, dip in a sponge, and wash the wall from the bottom up. Work in overlapping sections, so you don't miss any spots. Rinse with fresh water, and the wall will have a new lease on life. Don't forget to wear rubber gloves to protect your skin from the ammonia, and open a window or two for ventilation.

Stocking Up 🏠 TEXTURED WALL FINISHES are popular because they hide the seams and imperfections in the drywall. But the rough surface can shred a sponge to bits in no time flat, leaving particles and pieces clinging to the wall until you pick them off, one by one. So instead of using a sponge to clean textured walls, try a balled-up handful of old nylon panty hose or stockings. Either one will clean up the nooks and crannies without falling apart.

Reach New Heights 🏠 TO REMOVE DUST AND COBWEBS from your walls—especially where the walls meet the ceiling—put a clean paint roller on the end of a roller extension pole. You'll reach the dusty areas easily, and will be able to roll the webs away in a flash. If you don't have painting supplies on hand, simply attach an old T-shirt to the end of a broom to snag the dust.

Super Shortcuts

Vinyl won't absorb water, but moisture can seep into the seams when you're washing a vinyl-covered wall, causing the wall covering to lift, peel, or pucker. To prevent problems, use soapy foam instead of soapy water to clean it. Just whip ¼ cup of dishwashing liquid and 1 cup of warm water with a hand mixer until the mixture is the consistency of whipped cream. Dip a sponge or cloth into the foam and rub the dirt right off the wall. Wipe the foam off with a barely damp cloth or sponge, and the seams will stay nice and tight.

CHAPTER 5

FURNITURE
Fixers

Let's face it, most folks' furniture takes a beating over the years, resulting in everything from water rings and scratches on wood to spots and stains on upholstery. But good-quality furniture costs a pretty penny these days, so before you haul your flawed furnishings to the curb and head out to buy replacements, take a gander at this chapter. Follow my advice for fixin' furniture and you could have your table, sofa, or chair looking good as new again!

Black Ring Remedy

Erasing white water rings from wood furniture is hard enough. But if you see a dreaded black ring on your wooden furniture, you have a real problem. Black rings come from water that worked its way into the wood. And you have to fix the problem ASAP or risk living with that terrible stain forever. This remedy requires a lot of elbow grease, but it'll be worth it once you see that the stain is gone for good.

DIRECTIONS: Sand away the finish over the black ring as lightly as possible. Once it's gone, give the ring itself a light sanding. If it's stubborn, put on rubber gloves and bleach the wood. When the stain has faded enough to match the original wood color, rinse off the bleach and neutralize the wood with vinegar. Then apply matching stain to the bleached areas, and fill any dents or scratches with the wax filler. Apply a few light coats of varnish, using the steel wool to gently blend it in. Finish with the furniture wax, and you're done!

Sandpaper
Rubber gloves
Chlorine bleach
White vinegar
Wood stain (to match the original color)
Wax-filler stick
Varnish
Nonsoapy steel wool
Furniture wax

White Ring Out 🔲 TO GET RID OF a white ring on wooden furniture, mix together some baking soda and white non-gel toothpaste (proportions aren't important). Then use a damp cloth to rub the mixture into the stain, working with the grain. When the ring has disappeared, wash the area with oil soap, and apply a good furniture wax.

That's Brilliant!

When you reach for a soft cloth to clean your furniture, don't just dip into the ragbag and use whatever you grab first. Look for a cloth that's 100 percent cotton with a soft, lofty texture. Old cotton T-shirts work like a charm (just steer clear of any printed-on letters or decorations). Cloth diapers, flannel shirts or pajamas, and cotton socks will do the trick, too. Whatever you do, avoid synthetic fabrics, especially nylon, rayon, and polyester. They don't absorb well, and could even scratch your furniture.

Clean and Shine Furniture Fixer

> ¼ cup of boiled linseed oil
> ⅛ cup of whiskey
> ⅛ cup of white vinegar

Grandma Putt always made sure her furniture was clean as a whistle. Fingerprints, dust, and grime ran for cover when they saw her coming, especially because she used her own furniture polish. When it comes to making fine furniture look its elegant best, this easy-to-make polish gets the Grandma Putt Seal of Approval!

DIRECTIONS: Mix the ingredients in a glass jar with a tight-fitting lid, and wipe the mixture onto your wooden furniture with a soft, clean cotton cloth. Then buff with a second cloth. Cap any leftovers tightly, and store at room temperature.

Varnishing Vignette 🏠 WHEN IT COMES TO CLEANING varnished furniture, everyday dusting can be done with a soft, dry cloth. But your duties don't stop there. Varnished finishes need big-time protection. So polish them regularly with furniture wax, and keep liquids and foods away. That means if your varnished furniture gets a lot of use, you need to protect the finish with tablecloths, place mats, and coasters! To wax varnished furniture, use a soft cloth and work slowly, one small area at a time. Excess wax is enemy No. 1 for varnish, so make sure you buff each area well with a dry cloth. To keep wax from building up, polish sparingly—once every couple of months at most. In between sessions, you can always re-buff the surface to keep that shine going.

Super Shortcuts

✂ The next time you're tempted to pull out the furniture polish, try this all-natural alternative instead. Combine 2 tablespoons of lemon juice, 10 drops of lemon oil, and 4 drops of olive oil. Dip a soft, clean cloth into the mixture, and apply it gently to your furniture in a circular motion.

Fool-the-Eye Fixer for Fake Wood

You'd love to have a beautiful front door, but you simply can't afford an oak masterpiece. Don't give up hope—try this fake-out fixer that'll give your door the look of real wood without having to shell out the big bucks.

DIRECTIONS: Prepare the door by masking any hardware and filling holes and dents with Spackle. Evenly coat the surface with primer, and let it dry. Pull on a pair of rubber gloves, then measure the remaining ingredients into a sturdy bucket and stir them together. Using the paintbrush, apply a thin, smooth coat of the mixture on the door, then use the stiff brush to apply the next coat, creating long, curving lines that mimic wood grain. Keep working until you get the perfect look for you and your door!

*Available online and at woodworking-supply stores.

Spackle®
Primer
3 cups of mineral spirits
2 cups of satin-finish, oil-based varnish (not polyurethane)
1 cup of boiled linseed oil
3 tbsp. of burnt umber Japan color*
2 tbsp. of raw umber Japan color*
½ tsp. of Japan Drier*
Paintbrush
Stiff brush

Practice Makes Perfect 🏠 IF YOU'RE GOING for the fake wood look, practice your technique before you take the plunge. Paint the Fool-the-Eye Fixer (above) on a piece of scrap wood, varying your top coat technique until you're happy with the effect.

GRANDMA'S OLD-TIME TIPS

If there was one thing Grandma taught me, it was that while milk may do the body good, it has the opposite effect on wooden furniture. When you spill a glass of milk, wipe it up immediately. If you're stuck with a lingering stain, put a few drops of ammonia on a dampened cloth, and rub the spot gently. The milk should come right out.

Garden-Fresh Furniture Colorizer

When you're fixin' to "color up" unfinished wood, don't head to the paint store. Instead, hightail it to your supermarket and pick up some colorful produce. Different fruits and veggies are as useful for their colors as they are for their taste. Use this simple recipe to make your own stain.

> 1 to 2 cups of fruits, nut husks, or vegetables
> 4 cups of water
> ½ tsp. of alum*

DIRECTIONS: Put the produce in a pan with the water. Simmer for an hour, adding more water as it evaporates. Let it cool to room temperature. Stir in the alum as a fixative, brush the stain onto your wooden object, and let it dry overnight. Recoat, if you'd like, until the wood reaches your desired color density. Not sure what to buy to get the best color? For blue, choose blueberries, chestnuts, and red cabbage leaves; for brown, use walnut husks; for green, use spinach; for gold, use Golden Delicious apple peels; for orange, use yellow onion skins; for purple, use blackberries or purple grapes; for red, use beets, cranberries, red raspberries, and red onion skins; for yellow, use shredded carrots and lemon peels.

*Available in the spice aisle of your supermarket.

The Drinks Are On Us 🪑 OR, I SHOULD SAY, on the table! When you spill beer, white wine, or other alcoholic beverages on wooden furniture, wipe up the spill immediately, and rub the spot vigorously with the palm of your hand. Then dip a soft cloth in a little furniture polish, rub the stain gently, and wipe with a dry cloth.

Q *I have some small wooden items in my home that I would like to stain. A friend suggested that I use food coloring. This seems a bit odd to me. Does it really work?*

A Food coloring makes a great stain for planters, birdhouses, or anything made from unfinished wood (white pine absorbs color best). Mix 1 part food coloring with 6 parts water. Saturate the wood, wait five minutes, and wipe with a soft cloth. Let the piece dry overnight, then wipe again.

Grandma Putt's Furniture Polish

2 oz. of beeswax*
⅝ cup of turpentine**
Very hot (almost boiling) water

Most commercial furniture polishes are made with silicone, and these products work just fine on newer wood. But they give antique pieces an aura that doesn't look natural at all. Grandma used this homemade formula to polish her treasured tables and cabinets, and I still use it on my heirloom pieces.

DIRECTIONS: Coarsely grate the wax, and put it in a glass jar that has a tight-fitting lid (a glass mayonnaise jar is perfect). Add the turpentine, and screw the lid loosely on the jar. Stand it in a heat-proof bowl, and pour the hot water into the bowl so that it comes to or just above the level of the wax. Let the jar sit in the water until the wax has melted. Then remove the jar, tighten the lid, and shake the jar gently until a paste forms. Let the mixture cool, then pour it into a wide-necked jar (like a clamp-top canning jar) for storage. If the polish hardens, soften it up again by standing the jar in warm water. To use the polish, rub it onto the wood with a soft, clean cotton cloth, and buff with a second cloth.

*Available at craft-supply stores.
**Do not use mineral spirits or any other turpentine substitute.

Same Old Routine 🏠 IF YOU REGULARLY POLISH your furniture, make sure you use the same kind of polish on the same piece of furniture every single time. Why? Because mixing and matching wood-polishing products will leave your furniture looking hazy and streaked, and no matter how often you clean or dust the piece, it'll never look right.

That's Brilliant!

Wooden furniture comes with a protective finish that's not meant to be polished with oily products. So if you want to give your wooden furniture a little extra pizzazz once or twice a year, skip the oils and simply buff it with a small amount of beeswax on a soft cloth, going with the wood grain. That's all there is to it!

Heavy-Duty Leather Cleaner

> 1 part cream of tartar
>
> 1 part fresh lemon juice

If your favorite leather couch gets stained, don't hit the panic button. Use this paste to clean up dirty areas. It works great on heavily soiled spots like the front edge of your couch, the arms, or nasty "head prints" left by residue from hair products.

DIRECTIONS: Mix the ingredients in a small bowl, and work the paste into the soiled areas with a soft, dry cloth. Wipe it off with a damp cloth. If you still see dirt, reapply the paste and let it sit for a few hours to soak up the stain before you wipe it away. This potion won't keep, so make only enough to do the job at hand, then whip up a new batch next time the couch looks crummy.

Lather Your Leather REMOVE THE SURFACE DIRT from leather furniture, handbags, or clothes by rubbing it with just the lather from a bar of gentle, moisturizing hand soap. Dab the suds onto a soft, damp cloth, and rub the dirty spots on the leather. Wipe the lather away with a clean, damp cloth, and buff the area with a dry towel. And remember, use as little water as possible to make the lather, or you'll risk making water spots while you work.

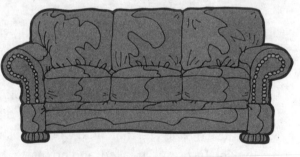

PINCHING PENNIES

Yikes! A stick of chewing gum opened up inside your leather handbag and now it's gumming up the works! Not to worry—unless you're looking for an excuse to get a new purse, you'll be able to remove that gum lickety-split. Just fill a ziplock plastic bag with ice cubes, and rub it on the gum until it's hard enough to scrape off with your fingernail or the edge of a credit card. Continue the cold treatment until most of the gum is gone, then heat the residue with a hair dryer and rub any remaining bits off with a soft, dry cloth.

Light-Wood Scratch Concealer

2 tsp. of fresh lemon juice

1 tsp. of cool tap water

1 tsp. of virgin olive oil (the darker the better)

Iodine (optional)

A few scratches can add a certain charm and character to vintage country furniture, but with fine antiques or new pieces, scratches and scrapes are a whole different story. Although this marvelous mixer won't repair gouges, it is one of the most effective formulas I've found for camouflaging scratches on light-toned wood.

DIRECTIONS: Pour all of the ingredients into a jar and mix well. If the color of the solution is too light, add iodine, one drop at a time, until you have the right shade. Dip a soft cloth into the liquid, and wipe it onto the scratch. Wait until it dries (probably about five minutes), then buff with a soft, clean cloth. Follow up with a coat of your regular furniture polish.

Crayon It In 🏠 THE NEXT TIME YOU FIND a minor scratch in a piece of wooden furniture, color the mark gone. Find a crayon in a color that matches the tone of the wood, and rub the wax onto the blemish. Then use your fingertip to blend in the color and smooth out the surface. The scratch will disappear, and no one will be the wiser.

Q *Even though I keep telling folks to use coasters or place mats, they keep putting their drinks and dishes directly on my wooden table. Is there any way I can make the marks left behind by their carelessness disappear?*

A Probably no table in history has ever gone through life without a scar or two. The good news is that light marks and scratches often vamoose if you rub them with a little paste wax. And if that doesn't do the trick, turn to a stronger solution: shoe polish. Just apply the polish directly to the scratch and watch it vanish!

Lovely Leather Cleaner 'n' Conditioner

2 parts linseed oil (not boiled)
1 part white vinegar

This simple solution will clean leather and keep it smooth and supple, adding a gentle shine that also repels water and stains. You'll need a glass jar with a tight-fitting lid; an empty short, wide-mouthed salsa jar will hold plenty of this perfect potion.

DIRECTIONS: Pour the linseed oil and vinegar into the jar, screw on the lid, and shake well before using. To apply, wipe it lightly over your shoes, sofa, or other leather items, and let it soak in for about an hour. Buff the leather with a soft, dry cloth to remove any excess and bring out the shine. Leftovers will keep indefinitely, so screw the lid back on and store the jar in a cool, dark cabinet.

Cure for Water Spots

WHEN WATER SOAKS into leather items (and leaves a distinct spot behind), cure the problem by giving it more water. That's right—moisten the spot and the area around it with a damp cloth, and blow it dry with your hair dryer turned to the cool setting. Then rub leather conditioner over the area, and the spot will blend right in with the rest of the finish.

Change Is Good

UNLIKE UPHOLSTERED FURNITURE, which you can buy in assorted colors and patterns, a piece of leather furniture is usually made from just one basic color (though there are some vibrant red and blue leathers out there). But even if you have basic black, brown, or tan leather furniture, it doesn't have to be boring. Accessorize your leather sofa with colorful pillows and throw blankets, and switch them from time to time for a whole new look.

That's Brilliant!

Use a dab of olive oil or walnut oil to help hide scratches and restore suppleness to smooth leather items. Rub a small amount of either oil in with a soft, clean cloth, wait an hour or so to let it soak in, then buff the surface to remove any excess. Try this in an inconspicuous spot first to make sure it doesn't discolor the leather.

Oiled Furniture Formula

Grandma Putt knew that the worst thing you can do to a piece of furniture with an oil finish is to treat it with furniture polish or wax. For routine cleaning, she just wiped away the dust with a soft cotton cloth. Then, every few months, she treated it to this formula.

> 2 cups of boiled linseed oil
>
> 2 cups of gum turpentine
>
> ¾ cup of white vinegar

DIRECTIONS: Combine all of the ingredients in a small bucket. Then dip a sponge into the solution, and gently wipe the surface of the furniture. (Wear gloves—this stuff will irritate even the toughest skin!) Let the formula stand for five minutes or so to loosen any tough dirt. Then wipe away the excess with a soft, clean cloth and buff to a shine with another soft cloth. (Make sure you get all of the formula off the wood, or you could wind up with a gummy residue.) Wash out the sponge and your gloves with hot, soapy water.

From Start to Finish ⬆ THE TYPE OF FINISH dictates how you care for wooden furniture. But how can you tell what type of finish you've got? It's easy—just try a simple test. Drop a little boiled linseed oil onto the surface. If the oil soaks in, the wood has an oil finish. If it beads up, the wood has a hard finish.

Go for Glass ⬆ If you have an old wooden table that was made back before polyurethane and other durable finishes came along, protect that fine wood by covering the surface with a sheet of bevel-edged glass. Then keep your glass-topped table clean by spraying it with a solution of 2 tablespoons of ammonia per quart of water, and dry with a soft cotton cloth.

PINCHING PENNIES

A lot of today's wooden furniture is coated with a clear poly-urethane finish, which protects the wood and provides a long-lasting finish—so you don't need to spend money on polishes. Regular dusting will usually do the trick. For stubborn dirt, just wipe it off with a damp cloth or sponge, and then dry it with a soft, clean cloth. And don't worry—the moisture won't sink in past the poly finish.

Perfect Upholstery Shampoo

This fast and easy formula will get your upholstered furniture clean quickly, gently, and for a fraction of the price you'd pay for store-bought upholstery shampoo.

> 1 cup of warm water
> ¼ cup of dishwashing liquid

DIRECTIONS: Combine the ingredients in a bowl. Then, using a wire whisk or handheld electric mixer, whip the solution until dry suds start to form. With a soft cloth or brush, slowly massage the dry suds into an inconspicuous spot on the upholstery and let it dry. If you see no change in color or texture, keep cleaning. Let the furniture dry, then wipe with a soft cloth dipped in warm water and wrung dry. Make sure the fabric is completely dry before anyone sits on it. You can speed up drying time by opening some windows or using a fan to blow-dry the fabric.

Doggone It! IF YOUR DOG LOVES LYING on the couch, your upholstered furniture will eventually smell just like he does. Deodorizing sprays don't work for long, so use this nifty trick to soak up the funk. Simply sprinkle the couch with baking soda and let it sit overnight (Fido will have to snooze elsewhere in the meantime!). It'll absorb the doggy aroma like magic, and your sofa will smell fresh and clean when you vacuum the baking soda away.

That's Brilliant!

Mineral spirits (paint thinner) work wonders on removing stains from upholstery. Pour some on a clean cloth and test in an inconspicuous area first to make sure the fabric color doesn't change. Then go ahead and dab some on the next time you find marks from lipstick, mascara, felt-tip markers, grease, or oil. Apply a small amount of the spirits with a clean cloth and rub, using just enough to get the stain out—you don't want to saturate the fabric. **Note:** Mineral spirits are flammable, so work in a well-ventilated room and away from all sources of heat.

"Salad Dressing" Scratch Remover

> 1 tsp. of fresh lemon juice
> 1 tsp. of vegetable oil

Grandma Putt loved to work with her hands. Whether she was weeding the garden or patching up a pair of overalls in the sewing room, she was always busy. And whenever a scratch showed up on her favorite wooden table, she mixed up a little "salad dressing" to make it vanish.

DIRECTIONS: Combine the lemon juice and oil in a small bowl. Dip a soft cloth into the mixture, rub the scratch, and it'll simply disappear! For deeper scratches, my Grandma had another great trick—she rubbed the nut meat of a black walnut or pecan into the scratch and let the "juice" sit for about half an hour to darken the wood. Once she polished the table, the scratch was invisible!

Easy Fix 🏠 IF ANY OF YOUR dark-colored wooden furniture gets scratched, just reach into the medicine chest, and grab the iodine. Then dip a cotton swab into the bottle, and dab that ding away. **Note:** Before you try this trick, test it first in a hidden spot to make sure the color blends in nicely.

Wake Up and Smell the Coffee! 🏠 DON'T LET SCRATCHES on dark wood age your furniture before its time. Just save the last few sips of tomorrow morning's coffee. Let it cool, then use a cotton swab to dab it on scratches and nicks. Let it dry, and then repeat the process until you're satisfied with the color.

GRANDMA'S OLD-TIME TIPS

Grandma Putt stored her shoe polish in her bedroom, because (of course) that's where her shoes were. But she used it all over the house to patch up dings and scratches in her wooden floors and furniture. Natural-color polish is perfect for any light-colored wood like pine or pecan. For darker woods, use tan, brown, or oxblood—experiment in a hidden spot until you find the shade that works the best.

Shellac-Finish Shape-Up Solution

1 part boiled linseed oil
1 part mineral spirits

You don't see shellac finishes on newer furniture, but it was all the rage back when Grandma Putt was keeping house—and today it's hot stuff in vintage furniture stores. If you're lucky enough to have some shellacked treasures of your own, give them a coat of this fabulous formula once a year or so. In between coats, dust the pieces with a dry cloth or the dusting brush attachment of your vacuum cleaner. But don't ever clean the pieces with water, because moisture—even high humidity—tends to make shellac sticky.

DIRECTIONS: Mix the ingredients in a small bucket, then dip a sponge or soft cotton cloth into the solution, and rub it evenly over the wood surface. (Make sure you wear gloves.) Wipe away the excess with a soft, dry cloth. If it's been more than a year since you've cleaned the furniture—or if you've just acquired a piece that hasn't seen its fair share of tender loving care in a while—you may need to repeat the process to remove all of the dirt. When you're through, wash your gloves and cleaning cloths in hot, soapy water.

A Knack for Shellac 🏠 SHELLAC FINISHES ARE BEAUTIFUL, but the finish is extremely brittle. It scratches easily and tends to develop crazes, especially as it ages. You can have a shellacked piece refinished professionally, but why not see how well you can repair it yourself first? Shellac-filler sticks are available at hardware stores and online. And in my experience they're easy to use—just follow the instructions on the label for best results.

That's Brilliant!

While this tip isn't about furniture, it IS about caring for wood. There's no doubt about it: Wooden hangers are a lot easier on your clothes than the metal and plastic types. Granted, sometimes, the wood can snag fabric, but you can solve that problem by sanding down the offending hanger and giving it a coat of clear shellac.

Shine Up the Wood Elixir

Wooden furniture can really make a room feel warm and homey. And it's not hard at all to keep that beautiful wood looking its best. Here's a fantastic furniture polish that'll work on any kind of wooden furniture.

> ½ cup of boiled linseed oil
> ¼ cup of malt vinegar
> 1 tsp. of lemon or lavender oil

DIRECTIONS: Put the linseed oil and vinegar in a clean jar with a tight-fitting lid, and shake it vigorously. Stir in the lemon or lavender oil. Apply the polish to your furniture with a soft, clean cotton cloth, and buff with a second clean cloth. (In hot, humid weather, reverse the proportions, so that you're combining ¼ cup of linseed oil and ½ cup of vinegar, because damp heat tends to make oil, well, oilier.)

Oil's Well That Starts Well 🏠 HERE'S THE GOLDEN RULE for furniture with oil finishes: Never treat it with furniture polish or wax (see "From Start to Finish" on page 141 to determine what type of finish your furniture has). Instead, for regular cleaning, just wipe away dust with a soft cloth. Then, every few months, give your oil-finished pieces a dose of something specially designed for treating those surfaces.

Excess Polish Remover 🏠 IF YOU WENT A BIT overboard on the furniture polish during your cleaning session, you can use cornstarch to absorb the excess. Just pour a little cornstarch onto the wood, and rub with a soft cotton cloth. It'll take the cloudy film right off.

GRANDMA'S OLD-TIME TIPS

Without question, Grandma Putt always had the best recipes in her bag of tricks. And her recipe for furniture polish was no different. Just mix 1 teaspoon of lemon juice with 2 cups of olive or vegetable oil. Wipe it onto your wooden treasures with a soft cotton cloth, then buff it to a shine with a second cloth.

Simple Saddle Soap

It's not just for saddles! Saddle soap is also the bee's knees (as my Grandpa Putt used to say) for cleaning leather-covered furniture and home accessories, as well as leather car upholstery. You can buy commercial brands, but it's as easy as pie to make your own using this simple recipe.

> 2 tbsp. of beeswax*
> ¼ cup of white vinegar
> ⅛ cup of linseed oil (not boiled)
> ⅛ cup of vegetable-based dishwashing liquid (like Seventh Generation™ Free & Clear Natural Dish Liquid)

DIRECTIONS: Put the beeswax and vinegar in a small saucepan and heat on low until the wax has melted. In a small bowl, mix the linseed oil with the dishwashing liquid and add the solution to the wax mixture, stirring until all of the ingredients are thoroughly blended. Pour the mixture into a shallow, heat-resistant container with a lid (like a candy tin or a glass jar that once held shoe cream). Once the soap has cooled and solidified, apply it to the leather using a clean, damp cloth, and then wipe it off using a second clean, damp cloth. **Note:** Always follow this treatment with a high-quality leather conditioner.

*Available at craft-supply stores.

Mildew Removal 🏠 HANDBAGS AND LEATHER CLOTHES tend to attract mildew when they're stored in a closet, and pesky mold can even get into your leather furniture if it's up against a damp wall. To remove mildew, wipe the affected items with a mixture of 1 cup of rubbing alcohol and 1 cup of water. But don't soak the stain—a few good swipes are all you need. Dry the area with a soft, clean cloth, and aim a fan or hair dryer (set on cool) at it to make sure every bit of moisture is gone.

That's Brilliant!

Wet leather stiffens and shrinks as it dries, making once-lovely items hard as boards—and clothing a size or two smaller! You can stop the shrinkage by putting the wet gear on a few times as it's drying out. When the leather is dry and has been stretched to keep its original shape, it's easy to restore its softness with a leather conditioner.

Surefire Cabinet Shine-Up Formula

½ cup of linseed oil (not boiled)

½ cup of malt vinegar

1½ tsp. of lemon juice

Don't let the name fool you. This potent potion will add shine to any wooden surface in your house—not just cabinets—whether it's painted, varnished, or lacquered. (In fact, the glow will be so bright, you may need to put on your sunglasses when you walk into the room!)

DIRECTIONS: Combine the linseed oil and vinegar in a small jar or bowl. Add the lemon juice for a fresh scent. Apply the polish with a soft cotton cloth, adding a little elbow grease, and your cabinets will be the talk of the town (or, at least, of your house)!

Easy Dusting Does It ● DON'T BE TEMPTED to polish your wooden furniture frequently because waxes or sprays can leave a cloudy buildup. Instead, stick to frequent dusting to keep the wood looking good. Use a dry microfiber cleaning cloth or a chamois, which will glide over the surface without leaving lint or scratches behind. If the wood piece has carved or raised decorations, use a synthetic or lamb's wool duster to really get in the groove.

Super Shortcuts

✂ If your wooden furniture isn't coated with a clear protective finish, you'll need to polish it about once a month to maintain the gentle shine—or wax it about twice a year. Oil-based sprays are quick and easy to use and require less buffing than waxes, but waxes do a better job of protecting the wood. Whichever kind of polish you use, wipe it on the wood in the direction of the grain, and then buff the polish off with a soft, clean cloth to leave a gleaming finish. And be sparing with furniture polish because a little goes a long way.

Teatime Wood Stain

> 2 tea bags of black tea
> 1 qt. of water

Do you have an old wooden table hidden away because you don't like the finish? Take it out of hiding and revive its look with a spot of tea. Just remember, tea will permanently stain the wood.

DIRECTIONS: Place the tea bags in the water in a saucepan, and boil for 15 minutes. Let the tea cool, and then test the color on a hidden spot of the furniture. If you like the color, dip a soft cloth in the tea, and wipe it all over the clean furniture. Continue applying the tea until you are satisfied with the results, then let the piece dry before finishing the look with furniture polish.

Grease-Stain Removal GREASE STAINS can be tough to remove from a wooden table, but this technique always worked for Grandma Putt. Saturate the area with mineral spirits—not paint thinner, which could damage the finish (and smell up the house, to boot). Then put an old, clean cotton cloth over it to soak up the grease. (Be sure it's 100 percent cotton, because synthetics can't absorb worth beans.) You may have to repeat the procedure a couple of times, but I guarantee it'll send that grease packin' for good!

This Bud's for Your . . . Table? IF YOUR FAVORITE PIECE of oak furniture is looking a little dull and lifeless, perk it up with a nice cold brewski. That's right, simply mix 2 cups of beer, 2 teaspoons of sugar, and a small lump of melted beeswax in a suitably sized jar. Brush the mix on, let it dry, then buff it off with a clean, dry chamois. Your old oak piece will look better than ever—guaranteed!

PINCHING PENNIES

You've found a terrific table for next to nothing at a flea market. Just one problem: There's a thick buildup of polish on the wood. How do you get the stuff off? Simple! Just wipe it with a half-and-half solution of white vinegar and water, and rub it off immediately. Repeat the steps if you need to (but you probably won't).

Vibrant Vinyl Rejuvenator

If you have vinyl-covered furniture, then you know that all kinds of dirt and grime like to settle on the surfaces. But this fabulous— and fabulously simple—fixer makes it a snap to clean that crud off lickety-split.

DIRECTIONS: Mix the ingredients together, and wipe the solution onto the dirty spots. Then wipe with a soft, clean cloth or chamois. Your vinyl will be clean as a whistle and rarin' to steal the limelight in a 1950s sitcom.

Vinyl Shine 🏠 KEEP YOUR CAR'S VINYL upholstery looking its best with this three-step routine. Clean the seats, dashboard, and other vinyl coverings with a damp cloth dipped in baking soda. Follow up by washing the surfaces with a mild solution of dishwashing liquid and water. Rinse the vinyl thoroughly with warm water, and then step back because it'll shine like new!

Baby Your Vinyl 🏠 SURE, VINYL IS DURABLE, but you still have to treat it gently if you want it to last. Using abrasive brushes or harsh household cleaners can scratch the smooth surface of vinyl, making it rough and brittle. So save yourself the aggravation and stick to soft cloths, soft-bristled brushes, and mild concoctions (like baking soda) to clean your vinyl surfaces.

Super Shortcuts

✂ The key to maintaining vinyl upholstery is to keep it simple. For general upkeep, just damp-wipe vinyl surfaces with full-strength white vinegar. Then dip another cloth in warm water and wipe the vinegar off. Make this a part of your weekly cleaning routine, and the vinyl won't become brittle—even in spots that see lots of wear and tear, like headrests and armrests.

Vim and Vinegar Leather Lotion

¼ cup of white vinegar
½ cup of water
Saddle soap

Who doesn't love the look and feel of leather? Unless, of course, its finish is crudded up with a waxy buildup of dirt. This potent but perfectly safe fixer will get rid of that gunk in no time flat.

DIRECTIONS: Mix the vinegar and water, and rub the leather with the solution, using a soft sponge or cotton cloth. Follow up by washing it with saddle soap and water. (You can find saddle soap at tack shops, most feed and grain stores, and shoe-repair shops, or you can make your own. See Simple Saddle Soap on page 146.) Complete the job by buffing the leather with a soft cloth to bring out its natural shine.

Take a Powder 🏠 TO MAKE GREASE STAINS disappear from leather or suede, blot the area with a dry cloth, then sprinkle cornstarch or talcum powder on the spots. Let the powder sit for several hours to absorb the oil, and then brush it away with a clean, dry cloth or a clean, dry nailbrush.

Water-Stain White-Out 🏠 TO REMOVE A WHITE water spot from your best leather-topped end table, rub a bit of mayonnaise into it. Let the mayo sit for an hour or so, then wipe the area clean with a soft, dry cloth. The white will erase the white, and your leather tabletop will be as good as new.

That's Brilliant!

Skin and hair are full of body oils that rub off on upholstered and leather furniture, leaving a greasy residue that attracts dirt like a magnet. So if this tends to be a problem in your house, dig out those old doilies or other removable covers and put them to use, especially on the armrests. Then toss a throw over the back of the sofa to keep hair products and grease from soaking into your upholstery or leather.

Wake Up and Smell the Furniture Cleaner!

Are you tired of using store-bought products to clean your fine wooden furniture? If so, then this remarkable recipe has your name written all over it.

½ cup of Murphy® Oil Soap
2 drops of patchouli oil*
1 drop of cedar oil
¾ cup of water

DIRECTIONS: Mix the ingredients together in a handheld sprayer bottle. Then dust the wood with a clean, soft cotton cloth (pieces torn from old flannel sheets, shirts, or pajamas are perfect for this job). Spray all of the surfaces with the cleaner, and wipe with a fresh cloth.

*Available at health-food stores.

Pretty in Paint 🪑 DON'T EVER WAX OR POLISH painted furniture, and steer clear of oil and oil-treated cloths. The paint is doing the job for you by protecting the wood beneath it, so there's nothing more you need to do except dust with a soft, clean cloth. If the paint gets chipped, simply sand the spots and retouch them with matching paint. And if your painted furniture is looking like it's seen better days, opt for a fresh coat of paint over the entire piece.

Please Pass the Mayo 🪑 GOOD OLD MAYONNAISE can do a lot more around the house than just dress up sandwiches and potato salad. You can also use it to remove crayon marks from wooden furniture. Simply rub on the mayo, let it sit for a minute or so, then wipe the area with a damp cloth.

PINCHING PENNIES

Save yourself the time and money it takes to repair or replace scratched and dented wooden furniture by preventing the damage in the first place. Just pick up a pack of felt furniture protectors next time you're at a home-improvement or discount store. Then place the pads on the bottoms of lamps, vases, frames, and other decorative items that sit on top of the furniture. Want to save even more money? Cut protectors out of felt, and secure them into place with plain white glue or double-sided tape.

CHAPTER 6

FLOORS
a Go-Go

Floors are always underfoot, which means they bear the brunt of snow-covered boots, muddy shoes, and sandy flip-flops. The easiest way to keep your floors looking their best (whether carpeted or made of wood, tile, linoleum, or vinyl) is to have your family and guests remove their footwear when they enter the house. But even then, you're going to have to deal with scratches, spills, scuffs, and stains on a fairly regular basis. Read on for my solutions to help make your floors look flawless.

All-Around Super Cleaner

If you've fallen into the habit of buying a different cleaner for every kind of surface in your house—one for your tile floors, another for your kitchen counters, yet another for your bathroom fixtures—I have just two words of advice for you: Stop it! Instead, whip up a batch of this excellent elixir and use it to tackle all of your cleaning chores. It even works on carpets and area rugs.

> 2 cups of water
> ½ cup of finely grated castile soap
> 2 tsp. of washing soda*
> 2 drops of vanilla oil
> 2 drops of wintergreen oil

DIRECTIONS: Bring the water to a boil, then remove the pan from the heat, add the grated soap, and stir until the flakes are fully dissolved. Let the mixture cool to room temperature and pour it into a bowl. Add the washing soda and the essential oils, and mix well. **Note:** Before you use this cleaner on a carpet (especially an older one), test it for colorfastness.

*Available online and at some hardware stores.

Get Buffed IF YOUR WOOD FLOORS need some spiffing up, the best tool to use is a floor buffer. But if you don't have one and don't want to go through the hassle of renting one, try my favorite method: Wrap an old cotton towel around the bottom of a dry mop and use that to buff your floors. When I was younger, I'd put on a few pairs of old cotton socks, and "skate" up and down the lengths of the floorboards, with the grain. A word of warning, however, if you try this method: Watch out for splinters!

PINCHING PENNIES

Got a waxy buildup on your floor that you need to take care of before your party guests arrive? Try this cheap and easy way to do it. Simply scrub the buildup with a solution of 3 parts water to 1 part rubbing alcohol, and rinse thoroughly. (Make sure the room is well ventilated before you begin.)

Can-Do Carpet Shampoo

The carpet in a busy home gets walked on, all over, all day long. So perk up the high-traffic zones with a batch of this shampoo. You may never buy a bottle of the brand-name stuff again.

½ cup of powdered laundry detergent

1 tsp. of ammonia

4 cups of warm water

DIRECTIONS: Mix the ingredients in a bucket, stirring vigorously until you get a thick froth. Dampen a large sponge with the froth only, and rub the entire carpet lightly. You'll have to stir the solution frequently to keep the froth from settling. When the carpet has been cleaned, let it dry, then vacuum thoroughly.

Join the Japanese 🏠 ANYTHING COULD BE HIDING on the bottoms of your shoes, waiting for a chance to rub off on your rugs! Save your carpets from wear and tear by training your family and friends to take off those dirt collectors as soon as they step inside, just like they do in Japan. If you keep a shoe rack with a few pairs of one-size-fits-most spa slippers inside the door, folks will remember to trade their shoes for slippers. Once everyone picks up the habit, you won't be picking up nearly as much dirt.

Pass the Peanut Butter! 🏠 WAIT A MINUTE—what the heck is peanut butter doing with a bunch of tips about carpet cleaning? It may sound nutty, but peanut butter is just the ticket for removing chewing gum from carpeting. First, heat the gum with a blow dryer to get it soft and loosen it from the fibers. Then use a plastic spatula to scrape up as much of the wad as you can. Once most of the glop is gone, work a small dollop of peanut butter into the remaining gum, and let it sit for five minutes or so. Wipe it up with a damp sponge, then dab the spot clean with a mixture of water and a squirt of dishwashing liquid.

Super Shortcuts

✂ You can put the bleaching action of denture-cleaning tablets to work on carpet stains by dissolving a tablet in a cup of water and using that mixture to sponge the stain away. Just be sure to test it first in an inconspicuous spot to make sure it's safe for your carpet. Otherwise, you could end up with a stain that looks worse than it did when you started.

Clean and Clear Wax Stripper

Are you having trouble getting rid of old wax on your linoleum, vinyl, or tile floor? When you have a floor that requires regular waxing, gunk can build up over time and leave your floor yellow and grimy. Make it clean and clear with this floor-wax stripper.

> 2 cups of ammonia
> ½ cup of Spic and Span® powder cleaner
> 1 tsp. of rubbing alcohol
> 1 gal. of cold water

DIRECTIONS: Mix the ingredients in a bucket, then use fine steel wool and good ol' elbow grease to remove the dingy old floor wax. Follow up by mopping the floor with a clear-water rinse.

The Whole Ball of Wax 🏠 I WOULDN'T CALL CANDLE WAX a stain exactly—but I'd sure call it a mess when it drips all over your floor! The good news, though, is that wax is a snap to get off. First, scrape off as much as possible with a plastic scraper or an old credit card. Then put a few ice cubes into a plastic bag, and hold it on the remaining wax. The stuff will become crumbly, and you'll be able to wipe it away. Just make sure you soak up any moisture that escapes from the ice bag if you're cleaning wax from a wood floor, or soon you'll have real spots to contend with!

Sunscreen, Please! 🏠 JUST LIKE YOUR SKIN, your hardwood floors will change color from exposure to the sun. Over time, this can cause permanent damage. While you can't slather on the sunscreen to keep your floors safe, you can take these steps: First, use window treatments to shade your floors. Second, rotate area rugs and furniture at least twice a year so unexposed and exposed areas of the floor age evenly.

That's Brilliant!

Got a cigarette burn on your hardwood floor where someone missed the ashtray? Apply a paste of equal parts white vinegar and baking soda to the mark, and use a pencil eraser to rub it in. Wipe off the residue with a damp sponge, and let the spot dry thoroughly. Then use a wood-stain marker to color the spot until it blends right in.

Dynamic Duo

> 1 cup of baking soda
> 1 cup of hydrogen peroxide

Scuff marks, grease, and dirty buildup on your linoleum floor can drive you bonkers. So here's a simple way to get rid of all of the smudges and smears. You'll whip up batch after batch of this super solution because the powerful pair of ingredients makes it a spot-on cleaner. And the formula helps to preserve the color of the linoleum as well.

DIRECTIONS: Mix the ingredients until a paste forms, then scrub grime away using a wet sponge dipped in the paste. Follow up by mopping the entire floor with clean water.

Linoleum Milk Cleaner 🏠 TO MAKE A LINOLEUM FLOOR sparkle, pour a little skim milk on it, and spread the milk around with a mop. Wipe away any excess from cracks and crevices, and follow this treatment up with a warm-water rinse.

Nail Polish Scuff Buster 🏠 GET RID OF SCUFF MARKS on a linoleum floor by wiping them with a dab of nail polish remover. (Test this method on a hidden part of the floor first.) Even though the stuff is made to clean up your nails, it works plenty darn well on floors, too!

GRANDMA'S OLD-TIME TIPS

Back in Grandma's day, folks covered just about every floor in the house with linoleum. True linoleum was made of ground-up cork and wood dust, mixed with pigments and a binding agent such as linseed oil, all attached to a canvas or felt backing, and cut into sheets and squares. If you have genuine linoleum, you need to give it special treatment on cleaning day. Linoleum is damaged by very hot water and strong detergents like ammonia because they cause the binding materials to dissolve. So next time you're spiffing up linoleum floors, use lukewarm water instead, and don't let it sit on the floor too long before mopping it up.

Fabulous Floor Cleanup

If you're going to wax your floor, you probably don't need to strip any old wax off first because well-buffed, waxed floors won't have any residue that needs to come up. However, if you have a heavily soiled floor, a floor with improperly buffed wax buildup, or a floor with clear, water-based acrylic polish, it'll probably need a more thorough cleaning. Use this solution to get old grime, wax, or acrylic polish off your floor before you put down a new wax coat.

DIRECTIONS: Mix all of the ingredients together in a large bucket. Use a sponge or a mop to soak a small area of the floor. Let the mixture sit for five minutes to soften the old wax, then scrub it with a nonabrasive plastic scrubbie. Sponge up the cleanser, and rinse the area with clean water. Repeat the process for the rest of the floor as many times as you need to, until no buildup remains.

*Available online and at some hardware stores.

When It's Time to Strip

SINCE WAX IS SELF-CLEANING and applying new wax will dissolve the old layer, it's rarely necessary to strip the floor before waxing. But if the old wax is dirty or discolored, it's probably time to take it off. The secret to strip-free waxing is in the buffing. A vigorous buffing that leaves only a thin film on the floor will give the surface a good enough coating for a glossy finish. The best (and fastest) way to get the job done is to use a floor buffer. But if you don't have access to one, don't worry. Just follow my tips in "Get Buffed" on page 153.

Super Shortcuts

Here's an old trick to cut down on the time you spend cleaning your floors, especially the areas near an entryway. Use walk-off mats before and after each entrance. For maximum effect, the mats should be at least four strides long. The idea is that you "walk off" the dirt from your shoes before you enter the room.

Firm-Traction Floor Polish

Grandma kept her wood floors bright and shiny with this non-slip formula. (This recipe makes enough polish to cover roughly 144 square feet of floor area—a room that measures about 12 by 12 feet.)

> ½ cup of orange shellac
> 2 tbsp. of gum arabic*
> 2 tbsp. of turpentine
> 1 pint of denatured (not rubbing) alcohol

DIRECTIONS: Mix the shellac, gum arabic, and turpentine until the gum arabic is dissolved. Add the denatured alcohol, and store the polish in a glass jar with a tight-fitting lid. Apply the polish to the floor with a soft cotton cloth. Wait half an hour, then buff the area with a second soft cotton cloth.

*Available at hardware stores.

Groom Your Broom

GRANDMA PUTT HAD A TRICK for making a new natural-bristle broom last longer. Before she used the broom for the first time, she'd soak the bristles in hot salt water for half an hour; then she'd shake them out and let them dry thoroughly. She said it made the bristles tougher, and by golly, she was right!

Breaking News

GOT A BIG PILE of crumbs on your kitchen floor but your dustpan is nowhere to be found? Then here's some news you can use: Grab a small section of newspaper and wet the edge so that it sticks to the floor. Then sweep the mess onto the paper and toss the crumbs in the garbage bin. How's that for good news?

That's Brilliant!

One of the hazards of mopping the floor is that every so often someone (probably the mopper) knocks into the bucket, spilling some or all of the cleaning solution out on the floor. If this happens to you more often than you'd like to admit, you may want to switch to my favorite kind of bucket—a square one. I prefer square buckets to round ones because they're much harder to tip over. But that's not all, folks! Square buckets are also easier to pour out of, can easily fit a rectangular mop head, and if I need to clean higher up, the bucket's much more stable on a ladder platform.

Fleas Be Gone Floor Wipe

If Fido or Fluffy picked up some fleas and you're concerned about an infestation of the little buggers, don't panic. Just whip up a batch of this excellent elixir and give the tiny terrors what for.

4 lemons, thinly sliced
Cold water
1 gal. of hot water
1 tsp. of dishwashing liquid
5 drops of pennyroyal oil*

DIRECTIONS: Put the lemons into a medium saucepan, cover them with cold water, and simmer on low heat for one hour. Remove the pan from the heat, strain out the solids, and pour the liquid into a bucket with the hot water. Add the dishwashing liquid and pennyroyal oil and mix well. Apply the solution to the floor with a damp sponge mop. Let it dry, then rinse with a clean, damp sponge mop.

*Available at health-food stores.

Quick De-Greaser

DROPPED A PIECE OF PIZZA facedown on the family room carpet? Not a problem—just sprinkle some baking soda or cornstarch onto the greasy stain, let it soak up the oil overnight, and vacuum up the whole mess in the morning. This will take care of the grease, but if there's a lingering stain, use one of the carpet and rug stain busters in this chapter. Then, be sure to remind whoever let the slice slip to eat in the kitchen from now on!

PINCHING PENNIES

Smoke, pets, kids, and daily life can all contribute to making your carpet smell stale—or even worse. And unfortunately, pricey store-bought carpet fresheners will only mask the smell temporarily, not eliminate it. What you need is an odor neutralizer, and I've got just the ticket! Mix 1 part powdered borax with 2 parts cornmeal, and use a flour sifter to sprinkle it all over your carpet. Wait about an hour to let it do its thing, and vacuum it away.

Grime-Fightin' Floor Fixer

Are you ready for a terrific tonic that'll make your linoleum floor clean enough to eat off of? Then give it a once-over with this simple solution.

> ¼ cup of powdered dishwasher detergent
> 2 gal. of hot water

DIRECTIONS: Mix the detergent and water in a bucket, then scrub the floor thoroughly with the mixture, and rinse it clean. See? I told you your floor would be clean enough to eat off (though I wouldn't recommend it!).

Linoleum Floor Cleaner ⬆ LIKE A LOT OF THINGS Grandma once had in her house, linoleum floors have staged a big-time comeback. If you have this classic covering underfoot in your kitchen (or any other room), keep it fresh and clean by mopping the floor once a month with a solution made from ½ cup of apple cider vinegar per gallon of warm water. This stuff will cut right through grease and dirt—and leave the air smelling sweet, too!

Scuff Mark Remover ⬆ TO GET RID OF black scuff marks left by careless shoe wearers on any kind of flooring, rub the spots with a paste made from 3 parts baking soda and 1 part water. Your floors will shine again in no time.

Super Shortcuts

✂ The formula you use to clean your kitchen floor depends on the kind of floor you have. You can't use the same preparation on wood as you can on vinyl—that would be like using nail polish remover to brush your teeth! So save yourself time and aggravation when you want to get old-fashioned linoleum good and clean. Just add a little dishwashing liquid to a bucket of warm water, and use it to mop the floor. To protect the linoleum from those dreaded scuffs and scratches, add a tablespoon of baby oil to the mop water. That's all there is to it!

How Dry It Is Carpet Bath

When your carpet starts looking a little dingy, don't bother renting a steam cleaner. Try dry-cleaning it instead. It won't do quite as thorough of a job as the deep-cleaning action of a steamer, but it'll take care of surface dirt so well that you'll appreciate the savings. And the best part is, you can make your own dry carpet cleaner with this simple recipe.

2 cups of baking soda
½ cup of cornstarch
4 to 5 crumbled bay leaves
1 tbsp. of ground cloves

DIRECTIONS: Mix all of the ingredients together, and store the mixture in a shaker jar. (An empty carpet cleaner can is perfect.) To use, dust the carpet thoroughly with the powder, wait at least one hour, then vacuum away.

Blot with Rubbing Alcohol ⬆ TO QUICKLY REMOVE most carpet stains, pour a little rubbing alcohol onto a dry white cloth, gently blot the spot, and wipe the stain away. Just be sure to test this trick first on a hidden area of your carpet to make sure it doesn't lift the color along with the stain.

Mix Your Drinks ⬆ WHEN A GLASS OF RED WINE gets a little bit tipsy—uh-oh, there it goes, right on the rug!—pour a bit of club soda on the stain, and let it go to town. After the soda has fizzed away for about 30 seconds, blot the stain and the soda with paper towels. Repeat if necessary.

Q *For some reason, I always go into a coughing fit whenever I sprinkle powdered cleaner on my carpet. Why is that, and is there any way to avoid it—short of not using powdered cleaners at all?*

A Baking soda, borax, and other powdery odor soaker-uppers can irritate your lungs and bring on a coughing fit. So play it safe by wearing a disposable dust-filter mask whenever you're using a mixture containing one of these substances. And to be on the even safer side, keep your kids and pets out of the room until well after you've vacuumed up the powder.

Lazy Man's Wax Substitute

White non-gel toothpaste
1 cup of white vinegar
1 gal. of water

If you're in no mood to wax your linoleum floor, but you want it to look like you did, use this recipe to buff it clean. Your floor will be so clean and shiny, no one will know that you didn't just put a fresh coat of wax on it. Come to think of it, with this potion on hand, why bother waxing at all?

DIRECTIONS: Squeeze some toothpaste onto an old toothbrush or a dry cloth, and use the paste to wipe away any scuff marks. Then mix the vinegar and water in a bucket, and use it to rinse the whole floor clean. And if you're ready to snap out of your lazy funk, wrap a towel around each foot and scoot yourself around the floor. That way, you can buff the floor to a high shine—and get a little workout in at the same time.

Bye-Bye, Buildup 🏠 HERE'S GOOD NEWS: If you (or previous owners) have waxed your floor and you're noticing a buildup, don't lose sleep over it. Mineral spirits will take that wax right off. Just put on your gloves, and go at the surface one small area at a time with mineral spirits and a cotton cloth. Be sure to wipe each spot clean before moving on to the next one. To really spiff up the floor, don't stop there. Go one more step and give the entire floor the once-over with a mix of warm water and mild dishwashing liquid. Then stand back and admire the results!

Fizzy Floor Freshener 🏠 TO REMOVE OLD grease stains from your floor, pour regular (not diet) cola on the stain and let it stay there for an hour. Then wipe it off—both the stain and the soft drink—and finish the job by giving the floor a good swabbing with a damp mop.

PINCHING PENNIES

You can avoid expensive repair work on your wood floors by protecting them from your furniture. To prevent chairs from scratching wood floors, give the bottom of each leg a light coat of wax about once a month. Either paraffin or candle wax will do the trick and keep chairs gliding smoothly.

Linger No Longer Odor Remover

2 cups of baking soda
10 drops of essential oil
(lavender is a good choice)

Your whole room will seem a lot cleaner once any lingering odor is gone from the carpeting. So whenever the area starts smelling a little stale, just sprinkle a generous dusting of this refresher on.

DIRECTIONS: Stir the ingredients well, then pour the mixture into a shaker jar, and let it sit for two days to dry completely (otherwise you'll risk staining the carpet). When it's ready to use, simply sprinkle the aromatic mixer across your carpet, let it sit for 30 minutes, and then vacuum it up.

Double the Doormats

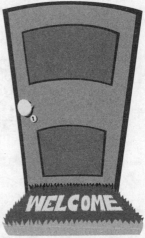

WHAT'S THE BEST way to keep dirt out of your carpets? Why, that's simple—don't bring it in in the first place! Make cleaning double quick by placing a doormat outside for folks to wipe their dirty shoes on, and another one inside, to catch the rest of the crud. Make that inside doormat a floor mat that's long enough to take three or four strides on, so it grabs even more dirt. Look for mats made from polypropylene, the best material for this dirty job—just shake them off frequently, or vacuum up the stuff they catch. And when they get really dingy, take 'em outside and hose 'em off.

That's Brilliant!

It's way easier to vacuum up dried mud from carpeting than it is to wipe up the wet stuff, so start by blotting up what you can with paper towels, then let the area dry. When the mud is nice and crusty, scrape off as much as possible with an old credit card or butter knife, and use your vacuum to suck up the clods and dust. Wiggle the rug fibers with your fingers to loosen any stubborn bits that are stuck deep down in the pile, and vacuum again.

Marvelous Mystery-Stain Remover

½ cup of white vinegar
2 tbsp. of salt
2 tbsp. of borax

Now where the heck did that come from? If you have a mystery stain fouling up your carpeting, give this marvelous mixer a try. Start with just the vinegar and salt; add the borax for extra-stubborn stains.

DIRECTIONS: Combine the vinegar and salt, and rub the mixture gently into the stain. Allow the paste to dry, then vacuum up the residue. If the stain won't budge, add the borax to the rub and follow the steps again.

Out with the Ink  AS SOON AS YOU SPOT an ink stain on your carpet, get to work quickly because the longer you wait, the harder it is to remove. First, blot up whatever you can with a paper towel. Next, sprinkle the stain with some cream of tartar, add a few drops of fresh lemon juice, and blot some more with a damp cloth, using a gentle wipe-and-lift motion. Be sure to fold the cloth as you work, so you're always using a clean section. Finally, vacuum up the cream of tartar, and you can write that stain off for good!

GRANDMA'S OLD-TIME TIPS

When my Grandma Putt wanted to make her wall-to-wall carpet smell extra good, she borrowed a magic ingredient from her spice rack—and added a pinch of cinnamon to her homemade carpet freshener. To spice things up yourself, mix ⅛ teaspoon of cinnamon with 1 cup of baking soda, sprinkle it lightly over your dry carpet, and wait about half an hour so the scent can soak in before you vacuum it up. The fine cinnamon dust can leave a stain if you grind it into the fibers, so don't walk on the mixture, and don't use this trick on white or pale carpets. For those, try light-colored powdered ginger instead of the darker-colored cinnamon.

Muddy Footprint Fixer

Lift muddy footprints and other stains out of your carpet with this quick and easy recipe. Just mix up the amount you need for the job at hand; one batch is plenty for a stain about the size of a single adult footprint.

> 2 tbsp. of dishwashing liquid
> ½ cup of warm water
> White vinegar

DIRECTIONS: Mix the dishwashing liquid and warm water gently, so that the solution isn't too sudsy, and dab it sparingly onto the stain using a sponge or a cleaning cloth; do not rub. Wait a few minutes, then lay a clean, dry cloth on the area, and blot up the soap solution. Now for the power boost! Pour a little white vinegar onto the stain, and blot again with a dry cloth. Then sponge on the soap solution again, blot with a cloth as before, and rinse by sponging with plain water. Blot with a cloth to remove as much moisture as possible, let the carpet dry, and then fluff the fibers with your fingers to make it look good as new!

Foam vs. Food YOU CAN MAKE RED WINE and other food stains on your carpeting or area rug vanish with the help of shaving cream. Just cover each stain with a shot of the cream. Then let the foam sit for 15 minutes or so, wipe it away with a damp sponge, and blot your carpet dry. If you don't have any shaving cream, you can try using a foaming bathroom cleaner—but test it first in a hidden spot to make sure it's safe for your carpet.

Super Shortcuts

Don't make a sour face at that carpet stain—just reach for the vinegar to make your rug look sweet again in no time flat! Here's how: Mix ⅓ cup of white vinegar with ⅔ cup of warm water, sponge it on the stain, and blot with a clean, dry cloth. You may need to repeat the process a few times, but it'll be well worth the effort.

No-Fuss Floor Cleaner

When it was time for my Grandma Putt to clean her floors, she whipped up a bucketful of this powerful potion that's so effective, we still use it at our house. It leaves even the dirtiest floor as clean as a whistle and as shiny as a new penny—without waxing!

> 2 cups of white vinegar
> ¼ cup of washing soda*
> 2 tbsp. of liquid castile soap
> 2 gal. of very hot water

DIRECTIONS: Put all of the ingredients into a bucket and stir to blend well. Mop the solution on the floor, and let it air-dry. Then stand back and admire the shine!

*Available online and at some hardware stores.

Cray-On, Scratch Off 🏠 IF YOU HAPPEN TO PUT a slight scratch or nick in a wood floor, don't worry. Just break out the crayons. Wax crayons work well because there are so many different colors, and you're almost certain to find a matching hue. Once you find a perfect match, rub the crayon lightly into the mark. Then buff the scratch (and the surrounding area of wood) with a soft, clean cloth. The scrape will blend right in.

Nail a Clear Solution 🏠 IT'S EASY TO TOUCH UP a surface scratch on your hardwood floor when it's only the finish that's marred and the nick hasn't penetrated the wood. All it takes is a dab of clear nail polish. Thin it by about half with lacquer thinner, then use the polish brush to paint the liquid on. Let it dry, then sand it with fine-grit sandpaper until the spot blends into the rest of the floor.

That's Brilliant!

Unless your attic has a finished floor, remember that the "floor" of your attic is actually the ceiling of your living room, dining room, and so on. Storing heavy items such as furniture or boxes of books can turn into a real headache if they come crashing through into the living space below. So before storing large items in the attic, lay some inexpensive lumber, such as pine boards or old shelves, across the ceiling joists, and place the heavy items on them.

Odor-Out Carpet Freshener

> 1 cup of crushed, dried herbs (such as lavender, rosemary, or southernwood)
> 1 tsp. of baking soda
> 1 tsp. of ground cinnamon
> 1 tsp. of ground cloves

Got a smell radiating from your carpet that you can't get rid of no matter what you try? Well, don't fret. Just mix up a batch of this fabulous fixer. It'll get right down into the fibers and root out that nasty aroma—instead of just covering it up with strong chemicals, as most commercial carpet fresheners do.

DIRECTIONS: Mix all of the ingredients together, sprinkle the combo over the stinky area, and work it into the fibers using a stiff brush or broom. Let it sit for half an hour or so, then vacuum. Now take a deep breath. Aaahh, the sweet smell of success! If you have any leftover mixture, store it in an airtight container.

Wipe 'Em Away, Baby! 🏠 FOR A REALLY FAST stain treatment, grab a baby wipe when something spills on your carpet, and simply wipe the mess away. Wipes are perfect for small, fresh stains because the alcohol and very gentle soap they contain remove most food stains or smears, and the cloth has just the right amount of moisture to remove a small stain without soaking the carpet.

Q *I've got a really stubborn stain on my carpet, and I've tried absolutely everything possible. I'm at my wit's end, and I don't know what to do. Is there anything else I can try?*

FAQ?

A To tackle a tough stain that just won't give up its grip, pour a small amount of hydrogen peroxide onto the spot. Let it fizz, then blot it up with a clean, dry cloth or paper towel. Repeat the steps until the stain is gone. Just be sure to dab a little hydrogen peroxide in a hidden spot before you start cleaning to make sure your carpet can take the treatment.

Scratch the Scorch Mixer

If your warm winter fire starts blazing like gangbusters and gets a little out of hand, you may end up with scorch marks on the carpet near the fireplace. Don't panic—and thank your lucky stars you didn't suffer worse damage—then use this easy-as-pie potion.

> 1 cup of white vinegar
> ½ cup of unscented talcum powder
> 2 medium onions, coarsely chopped

DIRECTIONS: Put all of the ingredients in a pan, bring the vinegar to a boil, and continue to boil for about three minutes. Remove the pan from the heat, let the mixture cool, and spread it over the scorch mark. (A soup ladle is just the ticket for this job.) Let it dry, then whisk it away with a stiff brush. Bye-bye, burn!

Salt for Soot 🏠 IF FIREPLACE SOOT sullies the carpet near your hearth, don't get all burned up about it. Simply sprinkle those black spots with plenty of salt, let it sit for at least half an hour to absorb the stain, and vacuum your troubles away.

Slice the Singe 🏠 TO GET RID OF minor burns from your carpeting, use tweezers to lift the singed fibers and cut them off using sharp scissors. Work carefully and slice away as little as possible. Otherwise you'll have a whole new problem to worry about—how to rearrange the furniture to cover up the hack job!

Q *My carpet looks great after I shampoo it, but afterward, it seems to get dirty faster than ever. Should I skip the DIY and have it professionally cleaned instead?*

A It's not your imagination—DIY shampooing can leave a sticky residue that attracts dirt like a magnet. But before you call in the pros, try this solution: Instead of filling the shampooer with soapy stuff, fill it with water and vinegar, using 1 cup of white vinegar for every 2½ gallons of water. Then "shampoo" your carpet, and let the vinegar lift out the old soap that's built up on the fibers. Empty the reservoir and refill it with warm water, and make another pass to be sure all of the soapy residue is gone.

Shine-On Hardwood Floor Formula

> 1 part vegetable oil
> 1 part white vinegar

If your hardwood floor sees a lot of action, it'll lose its luster pretty quickly. But that's no problem. Mix up a quick batch of this tonic to renew the shine on your hardwood floor in no time at all.

DIRECTIONS: Mix the ingredients in a handheld sprayer bottle, and then spritz the mixture on your floor, working in small sections at a time. Rub the solution in with a clean cloth, then buff the area with another cloth until your floor is clean and shiny.

High-Heel Horrors! NEVER, AND I MEAN never *ever*, wear high, thin heels on your hardwood floor. The spiky heels will make serious dents in the flooring faster than you can say "Jack Robinson." Believe me, all it takes is a single stroll across the boards. If you do get a gouge, you'll need to fill it to make it even with the rest of the floor. Start by rubbing the area with superfine steel wool, to make sure that it's smooth. Then, paint on layers of clear nail polish, allowing it to dry thoroughly between coats. Once the spot is filled in, the wood will shine through the polish, and you'll forget all about it.

PINCHING PENNIES

I love hardwood floors. They're beyond beautiful. But for all their beauty, they can be a real pain to care for. If you spill water on your hardwood, dry it immediately. If a spot remains, lightly sand it with fine sandpaper and then refinish the area. Or if you have a nasty cigarette burn, try a little steel wool moistened with soap and water. If the burn is serious, you may need to seek the advice of a professional. For heel scuffs, use a nonabrasive plastic scrubbie dampened with soap and water. Wipe the area dry with a soft, clean cloth. By keeping these tips and tricks in your arsenal, you'll save yourself some serious dough on repair costs!

Static-Stopping Solution

¼ cup of liquid fabric softener

2½ cups of water

When I was a kid, I loved to play practical jokes on friends and family by rubbing my feet on the carpet and giving 'em a good shock of static electricity. If dry winter air makes walking across your carpet a shocking experience, stop the static with this easy spray.

DIRECTIONS: Mix the ingredients in a handheld sprayer bottle, then lightly spritz the carpeting without soaking it. Stay off the carpet until it's completely dry. It'll be a no-zap zone for a week or two.

What a Shock!
🏠 IF YOU GET SHOCKED even when you're not walking across carpeting, take your shoes off! Walking barefoot and discharging the static electricity through your feet is just the ticket. The soles of your shoes (especially rubber soles) don't allow you to release the charge.

Keep Mud Off the Floors
🏠 DO YOUR KIDS WEAR a lot of extra gear when they go outside to play in the winter? Hang a shoe bag in the hall closet or mudroom to corral gloves, scarves, earmuffs, and other small outdoor accessories. Having your little ones stop right as they come indoors will keep them from leaving mud or melting snow all over the house. It may take a little patience until they get the hang of it, but once they know the drill, your floors will stay clean and dry.

That's Brilliant!

Has your four-legged buddy left a muddy paw print on your carpeting? You're probably tempted to tackle the muck right away—but don't! Make yourself wait until the mud dries, and use an old credit card to scrape off as much of the clump as you can. Then mix 2 teaspoons of mild laundry detergent in a cup of water and sponge it on the remaining spot. Rinse the area with a wet sponge, and blot dry with an absorbent towel.

Super Small-Scale Grease Cutter

2 tbsp. of white vinegar
½ tbsp. of washing soda*
¼ tsp. of liquid hand soap
2 cups of very hot water
½ tsp. of borax

For those times when you've got only a few greasy spots to clean—on your floor or any other hard surface—this is the fixer to reach for. It's every bit as potent as my True-Grit Grease Cutter (see page 173), but this recipe makes a smaller amount, so you can keep it in a handheld sprayer bottle for emergency cleanup jobs.

DIRECTIONS: Pour the vinegar, washing soda, hand soap, and water into a handheld sprayer bottle and shake to mix well. Spray the solution onto the greasy areas, and then buff it off with a soft, clean cloth. To make this mixture even stronger, add the borax to it.

*Available online and at some hardware stores.

A Butter Clean ⌂ WHOOPS! YOUR WELL-BUTTERED TOAST fell off your plate and landed butter-side down on the carpet. Don't panic. Instead, pour some cornstarch on the stain to absorb the grease. Let it dry, and vacuum up the residue.

Greasy Cement ⌂ GRANDPA PUTT ALWAYS KEPT a small sack of cement in his workshop for emergency patch-up jobs. If you have any leftover cement, then you've got a great grease cleaner-upper. To remove unsightly marks on your basement or garage floor, cover the spot with dry cement, wait 20 minutes or so, and sweep it up.

GRANDMA'S OLD-TIME TIPS

Whenever she needed to spiff up a cruddy kitchen floor, my Grandma Putt relied on a hardworking helper—Lestoil®. But she didn't stop there. Lestoil can tackle tough stains on carpeting, too. You'll find this old-time standby in the floor-cleaning products aisle of your local supermarket.

Terrific Terrazzo Treatment

1 cup of liquid
fabric softener
½ gal. of water

Terrazzo is a mixture of marble chips and concrete that's been polished to a high gloss. So when it comes time to clean a terrazzo floor, should you use a marble cleaner or one meant for concrete? The answer is: neither! To preserve the shine of a terrazzo floor, just wash it with this magnificent mixture. It'll clean and protect all in one easy job. And your terrazzo won't lose its gloss, either.

DIRECTIONS: Combine the fabric softener and water in a bucket, and wash the floor with the mixture. For a super-bright finish, add a splash of white vinegar to a bucket of fresh rinse water and mop it on the floor, then stand back and enjoy your reflection.

Floor It ⌂ WHAT'S THE BEST WAY to clean a ceramic tile floor? Use this couldn't-be-simpler cleaner to get the job done: Mix ¼ cup of white vinegar in a bucket of warm water, and use it to mop the floor. It'll shine like new.

Give Vinegar a Seal of Approval ⌂ YOU PROBABLY THINK of vinegar as the go-to cleaner for every room in your house. But if the grout on your kitchen floor isn't sealed, vinegar and other acidic cleaners should never be added to your mop bucket. Vinegar will clean the grout, no problem. But it'll also gradually dissolve it. The simple solution? Seal your grout.

That's Brilliant!

While good old-fashioned soap is fine for just about any cleaning problem, once in a while you come across something like a stone floor, where ordinary soap just won't do. It's not that it won't clean the floor, but soap will leave a residue on the stone that won't wash away easily, and the floor will end up looking kind of dull. To clean stone floors, I like to use a mild detergent, like dishwashing liquid, mixed in hot water. About a teaspoon to a gallon is all you need to do the trick. Mop the floor once with this solution, and once more with clear water for best results.

True-Grit Grease Cutter

Even in the best-kept homes, the kitchen floor can pick up more than its share of greasy dirt. This super-powered cleaner will cut right through that grease and grime while leaving your tile floor shiny and spotless.

¼ cup of washing soda*
¼ cup of white vinegar
2 tbsp. of dishwashing liquid
1 gal. of very warm water

DIRECTIONS: Pour all of the ingredients into a bucket and mix until the solution is sudsy. Then mop the floor as usual. **Note:** Don't use this cleaner on a waxed floor, because it could make the wax sticky and gunky.

*Available online and at some hardware stores.

More Floor Lore NOW THAT YOU KNOW all about how to clean different types of floors, it's time to think about tackling the toughest cleaning jobs. For example, get grease stains out of wood floors by immediately placing an ice cube on the stain. The grease will harden, and then you can scrape it off with a dull knife. Just be careful not to scratch your floor in the process, or else you'll have an entirely new problem to deal with.

Tar Remover TO REMOVE TAR from a floor, gently scrape up the excess with a dull knife, being careful not to scratch the floor. Then scrub any remaining gunk with a paste of baking soda and water. I don't even want to know how you got tar on your floor in the first place, but this will certainly take care of it.

GRANDMA'S OLD-TIME TIPS

Grandma Putt knew that talcum powder could do a lot more than soak up grease and moisture. For one thing, it silences your squeaky floorboards. Just sprinkle a little powder along the edges, and those boards will be as quiet as a mouse.

Weekly Wash for Vinyl

Vinyl floors have a built-in shine, so there's no need to use wax or polish when you're cleaning them, unless they've gotten dull from years of abuse. Before you apply a commercial product, though, try this powerful potion—it's a super-easy solution for keeping sheet vinyl or linoleum looking spiffy, and it costs just pennies to make. Plain water is enough to remove surface dirt, and the vinegar will cut through grease and give the floor a gentle shine. Best of all, there's no need to rinse!

DIRECTIONS: Mix the vinegar and water in a bucket, dip in your sponge mop, and have at it. Keep another bucket of clean water nearby to rinse your mop whenever it needs it. And don't worry about the vinegar smell—it'll disappear as the floor dries. **Note:** If you have a real linoleum floor, be sure to use only lukewarm water for washing the floor and rinsing the mop, and wipe the floor dry. Using hot water and/or letting water sit on a linoleum floor can damage the binding materials.

Down With Dingy ⌂ IF A VINYL FLOOR looks dirty no matter how hard you scrub it, you could be dealing with discolored floor wax or polish, or dirt that's worked its way under the glossy finish. To strip off the old wax, pour ammonia straight from the bottle onto the floor, and spread it over the surface with a damp sponge mop. The mop will start picking up dirt as soon as the ammonia dissolves the wax, so rinse it out frequently and keep mopping until the floor and mop are clean. Then rinse the floor with plain water, let it dry, and reapply wax if you like.

That's Brilliant!

To remove scuff marks from a vinyl floor, try a little dishwashing liquid on a damp sponge. If a troublesome area needs more cleaning power, use a nonabrasive plastic scrubbie instead of a sponge. You can also rub the spots with a pencil eraser, a dab of baking soda, or some white non-gel toothpaste to make the marks vanish.

CHAPTER 7

SUPER
Stain Removers

Stains, stains, go away—and DON'T come back another day! If you're tired of breathing in toxic fumes as you scrape, sponge, and scrub away stubborn stains, you can toss all of those commercial cleaners that are clogging up your cupboards right now. Just whip up some of these super-duper stain removers made from every-day household standbys. You'll soon be breathing a whole lot easier when those spots, blots, smudges, and smears do a disappearing act.

Beat the Brewski Stain Solution

Some folks look forward to kicking back with a cold beer at the end of a hard day. But if you get a little tipsy with the brewski, don't worry—treat the spot with this easy formula that's perfect for clothing and table linens.

DIRECTIONS: Mix the ingredients and use a sponge to dab the stain with the mixture, then toss the item in with the next load of laundry. For beer spills on carpet or upholstery, increase the vinegar in the mixture, using equal parts of vinegar and water. First sponge up as much of the spill as you can, then saturate the area with the vinegar solution. Blot the area with an absorbent towel.

Enjoy Your Dinner SPILLS WILL HAPPEN, especially at the family dinner table. But don't whip the tablecloth off in the middle of a meal just to tackle a stain. Go ahead and blot the spills up, then relax and enjoy the rest of your meal—you can take care of the problem later.

Q *I used my grandmother's old lace tablecloth for a dinner party, and my candles dripped onto the cloth. I was able to pick off some of the wax, but there's a lot more, down deep in the lace. Is there anything I can do to salvage this treasured heirloom?*

A There's no need to panic when candle wax drips onto your table linens, whether they're made of Grandma's lace or machine-washable cotton. Use this easy trick that'll make the wax disappear in a hurry: Set a box of frozen vegetables on the waxy spot for about 10 minutes, to harden it up. Then scrape off as much as you can with your fingernail or a dull knife. If the candle was colored, you're also going to have to treat the dye stain: Sandwich the stain between two layers of paper towels, and run a warm (not hot) iron over the top piece of toweling. The heat will melt any remaining wax. To keep the stain from spreading, replace the top and bottom paper towels as soon as they get greasy. Rub some dishwashing liquid into the spot, wait about 10 minutes so it can work, and then launder as usual.

Boot the Bloodstains Pretreater

> 1 cup of hydrogen peroxide
> 1 tsp. of powdered
>   laundry detergent

Because I grew up just like any other red-blooded American kid, I would often come home with some of my good old red blood on my clothes after a day of rough-and-tumble horseplay. But before Grandma Putt tossed my bloody duds into the washing machine, she got a head start on dissolving those stubborn bloodstains with this quick fix.

DIRECTIONS: Mix the peroxide and detergent in a small bowl, dip in a clean, dry sponge, and blot the mixture onto the bloodstain. Give it about 10 minutes to penetrate the stain, and then launder as usual. Just be sure to turn that dial to cold for the wash and rinse cycles! **Note:** Don't use peroxide on nylon.

Stain Remover 🔲 THE COMMERCIAL STAIN REMOVERS that we take for granted didn't exist in the days when Grandma Putt was laboring over load after load of laundry. Fortunately, she had a lot of spot busters that were just as effective as those fancy sprays and gels (if not more so). One of the best was—and still is—denture-cleaning tablets. To put these marvels to work on your clothes or table linens, just put the fabric in a container that's large enough to hold the soiled portion, fill it with warm water, and drop in two tablets. Leave the material in the solution until the marks disappear, then launder the item as usual.

GRANDMA'S OLD-TIME TIPS

There are lots of good ways to remove fresh bloodstains. Once those spots have dried, getting them out is trickier, but Grandma had great success with this routine: Soak the stained fabric in a solution of 2 tablespoons of ammonia per gallon of cold water until the spots have faded. Then, instead of using your regular laundry detergent, wash the item in cold water and dishwashing liquid.

Coffee Stain Solution

2 denture-cleaning tablets
½ cup of warm water

It's one of those universal laws: If you spill a cup of coffee, it will land on your white blouse (or shirt). Don't worry, just quickly spread the garment out in a sink or tub and pour this solution on it.

DIRECTIONS: Drop the tablets in a bowl containing the water, then pour the fizzing liquid over the stain. Let it soak for 30 minutes or so, then toss the top into the washing machine.

Join the Club Soda 🏠 KEEP COFFEE STAINS FROM RUINING cloth napkins, table linens, and other fabrics with this simple spray that'll neutralize acidic stains on the spot. Just fill a handheld sprayer bottle with club soda, spritz it on the stained fabric, and sponge the spot away with a soft cloth or sponge.

Coffee Talk 🏠 I LOVE THE SMELL of coffee in the morning! I just don't like the stains that the tasty beverage sometimes leaves on my shirts and pants. To get rid of coffee stains, soak the clothing in a sink full of vinegar and water overnight. In the morning, while you're enjoying another cuppa joe, lay the clothes out to dry in the sun. Just make sure that while you're sipping that second cup, you don't create even more stains!

Breakfast *On* Bed? 🏠 BREAKFAST IN BED is supposed to be a treat, but when the tea and toast get a little tipsy, put the power of borax to work. Drape the tea-stained sheet over the kitchen sink, sprinkle the stained spot with borax, and pour a pitcher full of hot water over the area.

PINCHING PENNIES Is your coffee mug stained? No, don't go out and buy yourself a new one! (Well, you can if you want, but it's not necessary.) Instead, soak that murky mug for 5 to 10 minutes in a solution of 1 tablespoon of bleach per gallon of water. This method works for tea stains, too. Your mug will come clean, and you won't have to waste a penny on new ceramic ware (or one of those big bucks coffee drinks).

Don't Sweat the Sweat Stains Solution

½ cup of bottled lemon juice, or 2 lemons
1 additional lemon
Plenty of sunshine

Don't let persistent perspiration stains come between you and a perfectly good white dress shirt. Before you buy a replacement, tackle those seemingly set-in yellow stains with this do-it-yourself recipe that combines lemon with sunshine, nature's best bleach. So far, nothing the detergent industry has come up with can top it!

DIRECTIONS: Pour the lemon juice, or squeeze two lemons, into a half-full washing machine tub of hot water. (If you're doing a full load, add twice as much lemon.) Soak your shirts for an hour or overnight, and then put them through the wash cycle. Add the juice of one more lemon to the rinse cycle, and hang out your shirts to dry in the sun. By the time they're dry, they'll be crisp, clean, and ready to wear.

Teamwork for Armpit Stains 🏠 FRESHEN UP THOSE DINGY underarm stains with a paste made of 1 tablespoon of borax and a little water. Just rub it in with an old toothbrush, and let it sit for about half an hour. Brush the crust away with a nonabrasive plastic scrubbie, and launder as usual. With a team of 20 mules working on it, that stain'll be history in no time at all!

Super Shortcuts

✂ If your clothes desperately need a bath when you're traveling, but you don't have extra duds on hand, just hop in the shower—with your clothes on! Wet yourself down thoroughly, and lather up your clothes with bar soap or shampoo. Slowly turn around in the spray, lifting the clothes away from your body so they get a thorough rinse. Peel off your clothes, wring them out, wrap them in a bath towel to soak up more moisture, and hang them over the shower-curtain rod to dry. Then it's time to get yourself clean.

Easy Brick Cleaner

Interior brick floors and walls are charming—until dust and dirt turn them into dingy old eyesores. Bring back the beauty of the brick with this quick-and-easy cleaner.

> 3 tbsp. of washing soda* or trisodium phosphate (TSP)
> 3 gal. of warm water

DIRECTIONS: Start by vacuuming the brick using a duster attachment. Then mix the washing soda or TSP into the water and use a small brush to apply the solution. Give the surface a good scrubbing, rinse the brick clean with warm water, and start enjoying your beautiful brick again! For stubborn stains, let the solution sit on the stain for 15 minutes or so before scrubbing and rinsing it off. **Note:** If your brick is more than 30 years old, it may be soft and can be damaged by vigorous scrubbing. Test a small area first, and if the brick crumbles, stick with frequent dusting.

*Available online and at some hardware stores.

Grease-Grabbin' Paste 🏠 WHEN GREASE FROM THAT nice juicy burger drips all over your shirt, reach for a terrific grease grabber—cornstarch! Mix it with enough water to make a paste, and slather it on the wrong side of the fabric. Brush it off once the paste is dry, then launder the garment as usual.

That's Brilliant!

When it comes to caring for table linens, grease stains are pretty darn common. And making them vanish is a cinch—as long as you pretreat the stains. It's essential, then, to inspect all linens for trouble spots before you toss them in the washing machine. Rub a squirt of dishwashing liquid into each spot, let the linens sit for about 10 minutes, and then launder them with an enzyme detergent to help dissolve the stains. Check the items again before you put them in the dryer because heat will set the stains. If you still see spots, repeat the treatment.

Fat 'n' Sassy Leather Waterproofer

> 2 parts beeswax*
> 1 part mutton fat**

If your shoes are made of leather, you're going to want to protect them when the nasty, wet weather comes around each year. Treat your hiking boots with this old-time conditioner, too. They'll stand up to anything the great outdoors can deliver and keep mud, grass, and other stains at bay.

DIRECTIONS: Melt the wax and fat together, stirring well. Rub the mixture onto your shoes or boots in the evening, and let them sit overnight. (Just keep 'em well out of reach of your dog or cat, who will be attracted to the fat and chew up your shoes in no time flat!) In the morning, buff the shoes with a soft cotton cloth. After this treatment, they'll be better prepared to handle whatever wet conditions you throw at them, as well as any surprise rainstorms.

*Available at craft-supply stores.
**Ask your butcher or meat department manager to order mutton fat, or substitute high-quality beef fat.

Old Salt DO YOUR LEATHER WINTER BOOTS or shoes have white stains left by snow-melting salt? You can blot the spots away with a simple solution of 1 part white vinegar and 1 part water. Repeat the treatment until the stains are gone, and then polish the footwear as usual. And while you're at it, treat your footwear with waterproof spray to make sure salt stains don't set in the next time you or your loved ones are out walking in a winter wonderland.

PINCHING PENNIES

Have you ever wondered if there's anything you can do with a banana peel before you throw it out (or compost it)? Well, there is. Turn that peel inside out and use it to shine up your smooth leather shoes. After a rubdown, buff the leather with a soft cotton cloth to work in the banana oil and bring out the shine.

Ground-In Spot Solution

1 part ammonia

1 part dishwashing liquid

1 part water

You can help solve most of your laundry issues by treating any ground-in spots before you throw the garment in the washing machine. After all, it would defeat the purpose if you didn't get your clothes clean when you washed them. This magical mixer is just the trick for stubborn stains on washable fabrics, including cotton, linen, polyester, and acrylics. Just spray it on to make grass stains, food stains, and other tough trouble spots disappear!

DIRECTIONS: Fill a handheld sprayer bottle with the ingredients, shake it up, and spray the mixture on the stains. Wait 10 to 15 minutes, and then launder as usual. This super solution works just as well, if not better, than store-bought pretreaters, so give it a try.

Bargain Alert 🏠 MAKING YOUR OWN LAUNDRY DETERGENT is a piece of cake. Just mix equal parts washing soda (available online and at some hardware stores) and borax. Add ½ cup of this mixture to each load of laundry. Even though you may use store-bought detergent for most loads, do several loads with this mixture at least once every month to get the detergent residue out of your clothes and give them a longer life. When it comes time for the rinse cycle, add ¼ cup of white vinegar to the load, and your clothes will come out just as soft as they do with a commercial fabric softener—and at a tiny fraction of the price!

GRANDMA'S OLD-TIME TIPS

Grandma knew exactly how to make any kind of stain scram. To get rid of organic, protein-based stains like milk, egg, and blood, make a paste of meat tenderizer mixed with a few drops of water. Work it into the stain, and then launder as usual. Tenderizers contain enzymes that break down proteins. If it works on a tough piece of meat, it'll work on a tough stain, too!

Hard-Surface Paint Stain Potion

½ cup of washing soda*
3 gal. of water

Old-fashioned washing soda is the key to making paint stains disappear from granite, slate, sandstone, concrete, and other stone surfaces. It's related to baking soda, but it's more caustic, so be sure to wear long rubber gloves when you work with the stuff.

DIRECTIONS: Mix the ingredients together in a bucket, dip a soft-bristled brush into the solution, and scrub away. Rinse thoroughly with fresh water, and let the spots air-dry. Washing soda works great on counters and vinyl floors, too, but it'll take off any protective wax along with the spots of paint. So you'll need to apply another coat of wax after the floor is dry.

*Available online and at some hardware stores.

Salt Away Stains 🏠 If you've got a stain on your marble floor, cover it with salt, then brush it away. Reapply and brush the salt off as many times as you need to until the stain is completely gone.

Fresh Paint Solution 🏠 DID YOUR NICE BLOUSE suffer a close encounter of the wet paint kind? No problem—if you treat the spot while it's still wet. Start by saturating the paint with warm water. Then rub in a few drops of gel dishwasher detergent, and rinse. But if the stain has dried, you may need to try this last resort to save your blouse: Spray it with oven cleaner. Make sure you do this in a well-ventilated area, let the garment sit for 15 minutes, then rinse it thoroughly.

That's Brilliant!

Concrete counters or walkways can go from spotless to dingy in no time flat. That's why I always keep a supply of this super-powered cleanser on hand. To make it, mix 2 cups of rock salt with 6 cups of sifted sawdust and 1½ cups of mineral oil. Store the mixture in a plastic garbage can with a tight-fitting lid. When the need arises, just scoop some out, sprinkle it on the surface, and sweep it up—along with a whole lot of dirt.

Ink-Eradicating Paste

You're furiously scribbling a note when suddenly the pen flies out of your hand and lands smack-dab on the carpet, leaving you with an inky mess. Don't panic—you can easily make the blotch disappear in a jiffy with this nifty tonic.

DIRECTIONS: Mix the ingredients into a paste, then after blotting at the stain to remove what you can, apply the paste to the spot. Blot the stain—don't rub it, or you'll drive the ink farther into the fibers—until the ink has disappeared. Then rinse the area with clear water, and soak up the water with an absorbent towel. You may also have to vacuum the area after it's completely dry to pick up any cream of tartar residue.

Tomato Ink Remover

WHEN I WAS A BOY, we had inkwells on our desks at school, and we did all of our lessons with old-fashioned fountain pens. Although I loved writing with those things, I was not, shall we say, the most careful kid on the block. I was forever coming home with ink stains on my shirt, and sometimes my pants, too! Fortunately, as usual, Grandma knew exactly what to do: She'd make me change into my play clothes, pronto. Then she'd lay the stained fabric on a flat surface and put a slice of raw tomato on each ink blotch. I'm not kidding! When the tomato had absorbed all the ink, she'd launder the garment as usual, and the ink was history.

Super Shortcuts

Don't cry over smeared ink—not even when it's leaked out of a ballpoint pen onto your favorite leather handbag or leather jacket. Instead, dip a soft cotton cloth in milk, and rub the marks away. Then wipe with a second cloth that's been dampened with clear, cool water. Presto—no more stain!

Melted-Crayon Miracle Mixer

> 1 cup of liquid laundry detergent
>
> 1 cup of liquid OxiClean® Laundry Stain Remover
>
> 1 cup of white vinegar
>
> 1 cup of Zout® Triple Enzyme Clean™ liquid laundry stain remover

Your little artist probably loves to scribble in coloring books all day. But it's a real "Oh, no!" moment when you find melted crayon all over your clothes—those smeary stains look like they'll be impossible to get rid of. That's when it's time to turn to this remarkable recipe.

DIRECTIONS: Start filling your washing machine, using the warmest water setting your clothes can tolerate, and add the ingredients as the water is running. When it's full, turn off the machine or set your soak cycle, add your clothes, and let them sit in the powerful solution for about 45 minutes. Then finish the cycle, and take a good look at your clothes. If the stains are all gone, toss them in the dryer. If there's still some crayon color, repeat the treatment.

Crayon Remover 🏠 HAS A YOUNG ARTIST been using your wallpaper as a canvas for coloring? Don't worry (as long as the paper is washable). Just brush a light coat of rubber cement onto the crayon marks, let it dry, then use your fingers to roll it—and the colorful wax—right off. Your wallpaper should be fine, with no lasting damage.

GRANDMA'S OLD-TIME TIPS

By the time commercial spray lubricants arrived on the scene, Grandma Putt had a passel of great-grandchildren. So when those budding Picassos and Georgia O'Keeffes filled her walls with crayon drawings, Grandma knew exactly what to do: She reached for her can of oil, sprayed the marks lightly, and wiped the wall clean. (Of course, she took a snapshot of the artwork first!)

Nix the Nicotine Buildup

If you have—or had—smokers in your house, your oil or acrylic paintings may be stained by a nicotine buildup. You can remove the grime by gently washing the surface with a sponge moistened with this soapy solution.

> ¼ cup of fresh or bottled lemon juice
> 2 to 3 drops of dishwashing liquid
> 2 cups of water

DIRECTIONS: Mix the ingredients together in a bucket, and keep a separate bucket of fresh water nearby so that you can rinse out the sponge after every few strokes. Then set to work sponging on the solution. Rinse it off with a clean, moist sponge, and that dull film will do a disappearing act.

"Hand"-y Help 🏠 YOU SHOULD ALWAYS KEEP a tub of degreasing hand cleaner in your laundry room because it's sheer genius when it comes to getting tough, greasy dirt out of clothes. Just rub it into the stains with your fingers or an old toothbrush, let it sit for about 15 minutes, and rinse the stains away. Old stains may need two treatments before they'll ease their grip, but ease their grip they will!

FAQ ?

Q *I have a great big family. So we go through stain-remover sticks like there's no tomorrow, and the cost really adds up! What else can I use to pretreat stains?*

A Stop busting your budget, and mix up a batch of my handy all-purpose stain remover whenever it's wash day (which just might be every day in your case!). Combine 1 part baking soda, 1 part hydrogen peroxide, and 1 part water, and scrub the paste into the stain with an old toothbrush. Let it sit for about half an hour, and then scrub again until the stain is gone. Prepare this mixture just before using it, so the peroxide doesn't lose its potency. This treatment takes just minutes to make and, best of all, it costs only a few pennies.

Paint Spill Disaster Diverter

> 2 tbsp. of dishwashing liquid
> 2 tbsp. of white vinegar
> 2 qts. of warm water

You're just about to finish painting the living room, when—OOPS!—down goes the paint bucket, splashing latex paint onto your carpet. Sponge the spill up with this tonic, then you can relax—after all, now you have nice freshly painted walls and a clean carpet!

DIRECTIONS: Mix the ingredients together in one bucket, then tackle the spill, rinsing your sponge between wipes in another bucket filled with clean water. The paint will lift out of your carpet in no time at all.

Sandpaper Scorch Remover 🏠 REMOVE A CIGARETTE or candle burn from a carpet by rubbing the mark with fine sandpaper. Work with a light, circular motion, and keep at it until the spot disappears. The scorch mark will soon be gone, and you'll hardly remember it was there in the first place.

Make It a Clean Sweep 🏠 IF YOUR CARPET'S LOOKING a bit grungy, try this trick before you scrub it. Start by sweeping it vigorously with a broom. This will make the nap stand up and loosen embedded dirt. Then vacuum thoroughly. Now cast a critical eye on your carpet. It may look so good that you'll decide it doesn't need any further cleaning!

That's Brilliant!

General carpet care falls under Common Sense 101—if you know how to use a vacuum, that's half the battle. But there's a bit more you should know in order to keep your carpets and rugs looking their best. If something spills, blot it up immediately. Soak up the excess stain, and then add a little white vinegar and scrub the area gently. If you want, you can also sprinkle a little baking soda over the carpet, then vacuum to eliminate lingering odors.

Pencil Mark Removal Potion

Using a pencil to outline the pattern on quilts or other craft projects can leave visible marks, especially if you use a soft, thick pencil. To erase all traces of your outlines on washable fabrics, try this potion—it does a real number on the marks, especially on white or light-colored cloth.

> ¼ cup of rubbing alcohol
>
> A few drops of dishwashing liquid
>
> ¾ cup of water

DIRECTIONS: Mix the ingredients in a handheld sprayer bottle and spray the potion onto the pencil marks. Wait about five minutes, and then rinse the marks away. Don't rub the fabric, and if you're not sure if your project is color-fast, try the spray on a leftover piece of fabric first.

Easel Does It IF YOU HAVE KIDS or grandkids, it'll probably happen: They'll grow tired of the coloring books you've given them and turn to the oh-so-inviting blank canvas before them—your walls! For pencil marks, wipe with a damp sponge and a little dishwashing liquid, and the marks will come off easily. Ink marks may take a bit longer to completely disappear. You can also try lightly scrubbing them with a paste of baking soda and water, then rinsing.

Chalk It Up to Creativity STOP CLEANING CRAYON MARKS off the walls, and give your artistic children and grandchildren their own wall space that they can color whenever they want. Just pick up a bucket of chalkboard paint at the hardware store, and tape off a large rectangular area that's low enough for little hands to reach. Brush on a few coats of the paint and then frame the area with decorative molding. Now all you have to do is hand over a pack of colorful chalk and let 'em rip—inside the frame only, of course.

Super Shortcuts

Did your artistic children or grandkids leave crayon marks on the wall? Try this neat trick: Squirt a generous amount of white non-gel toothpaste onto the marks, rub the area with a scrub brush, and rinse with a water-soaked sponge. That'll give those marks the old brush-off!

Pet Stain and Odor Eliminator

To neutralize odors and remove yellow or brownish urine stains on your carpet, mattress, or upholstery, give this potion a try. Test it in an inconspicuous spot (like under a seat cushion) first to make sure the peroxide doesn't bleach the color out of the fabric.

1 cup of white vinegar
1 cup of water
1 cup of baking soda
¼ cup of hydrogen peroxide
1 tsp. of dishwashing liquid

DIRECTIONS: After you blot up the urine, wet the stained area thoroughly with a mixture of vinegar and water. Blot it with paper towels until it's damp, not wet, and then sprinkle the baking soda liberally over the stained area. Finally, combine the peroxide and dishwashing liquid, and pour the mixture over the baking soda. Work it in with a scrub brush until the baking soda is dissolved and the mixture penetrates the fabric or carpet fibers. Allow it to dry, and then vacuum up the residue. Cat urine spots can be really stubborn, so if the stain reappears, repeat the treatment.

Climbing the Walls 🏠 WHEN YOUR CAT SPRAYS urine on the walls, start the cleanup by washing the spot several times with white vinegar. If the stain has penetrated the drywall, plaster, paneling, or brick, consider applying a clear, fast-drying sealer to keep the signature scent from perfuming your whole house. You'll find sealers appropriate for various types of wall surfaces at most home-improvement or hardware stores; just follow the directions on the label.

Urine can permanently stain wooden floors, so you need to act fast to prevent irreparable damage. But don't run out and buy an expensive rug cleaner. Instead, wipe up the area with wet paper towels, then wash it with undiluted white vinegar. Work quickly so the wood doesn't absorb the liquid, and dry the floor with fresh paper towels. Repeat the vinegar wash, dry the spot again, and your floor will be fresh and clean.

Red Wine Disappearing Act

> 1 cup of hydrogen peroxide
> 1 tsp. of dishwashing liquid

This mixture is hands down the absolute best way to make red wine stains disappear, whether they're on carpet, clothes, or furniture. It works like a son of a gun on fresh stains, old stains, and even set-in stains that have been through the washer and dryer. The only drawback is that you'll need to mix up a fresh batch whenever you need it because leftovers won't keep well.

DIRECTIONS: Pour the hydrogen peroxide into a small glass bowl and stir the dishwashing liquid into the peroxide. (If you're using the popular blue-colored dishwashing liquid, don't be surprised if it becomes clear as you stir.) Soak a clean cloth or sponge in the mixture, squeeze out enough liquid so that it's wet (but not sopping), and then blot the wine stain until it fades away. The peroxide in the mixture has a mild bleaching action, and although it usually won't be strong enough to lift the color out of fabric, test it in an inconspicuous area first just to make sure.

High Water 🏠 MY GRANDMA PUTT treated fruit juice stains on her tablecloth by holding the cloth over the sink while she poured boiling water through it. Wine is just fancy fruit juice, so put the teakettle on the stove and pour the steaming water on (and through) wine stains while they're still fresh. And get a helper to hold the cloth over the sink while you carefully pour.

That's Brilliant!

When a glass of wine gets tipsy on a tablecloth, immediately cover the stains with salt. Pour a generous amount onto the spots, covering them completely. The salt will soak up the wine and counteract the stains while you finish the meal. After your dinner guests have departed, shake off the salt, rub some dishwashing liquid on the spots, and launder the cloth in cold water.

Ring Around the Collar Remedy

Maybe we should call this remedy "black" magic, considering the way the inside of a collar gets so much dirtier than the rest of the shirt! So what's the solution? Try clearing things up with this homemade "ring cleaner."

> ¼ cup of **baking soda**
>
> 2 tbsp. of **white vinegar**

DIRECTIONS: Mix the baking soda and vinegar to form a paste. Use an old toothbrush to scrub it into the stained collar or cuffs, let it sit for about half an hour, and then launder as usual. No more ring around the collar!

Black Triangle = Red Flag ⬟ CARE LABELS ON CLOTHES are a great idea, but some of those symbols—like that solid black triangle with a line through it—can be tricky to figure out. FYI: That triangle is a big red flag that means "Do not bleach!" If there is no care label attached, avoid bleach whenever you wash silk, wool, or anything with spandex because the bleach can damage the fibers of those fabrics. And who wants saggy, baggy stretch pants?

GRANDMA'S OLD-TIME TIPS

Although red was my Grandma Putt's favorite color of lipstick, it's one of the hardest shades to get out when it's smeared on a collar or shirt. Why? Because a lipstick stain is part grease and part dye. So Grandma came up with a great way to handle this tough customer: Start by gently rubbing a bit of vegetable or mineral oil into the stain, letting it soak in for about 15 minutes. Then blot the spot with paper towels to remove as much oil and lipstick as possible. Next, sponge the stain with rubbing alcohol to lift the dye and cut the grease. Wait about five minutes for the alcohol to work its wonders, and then toss the item into the washing machine, using the cold-water setting. If any stain remains after washing, don't put the garment in the dryer—apply more alcohol, wait about five minutes, and then wash it again.

Straight-Shootin' Mold and Mildew Killer

> 1 cup of ammonia (not sudsy)
> ½ cup of white vinegar
> ¼ cup of baking soda
> 1 gal. of water

Mold and mildew can be two of the worst stain problems that a bathroom or kitchen (or any room for that matter) can encounter. Not only do they look unsightly, but they're spores that can be toxic to humans if ingested in any way. So knock mold and mildew out with this no-rinse, all-purpose spray formula.

DIRECTIONS: Mix the ingredients together in a bucket, pour the solution into a handheld sprayer bottle, and aim the spray right at the mold and mildew stains. Rub them out with a soft, clean cloth. Repeat the process until the stains are gone. There's no need to rinse!

Mildew Chaser 🏠 FABRIC THAT'S STORED in a garage or basement (or left there by accident) is a prime target for mildew. If these nasty spores have attacked your clothes, linens, or even favorite cleanup rags, try this formula: Paint the stains with lemon juice, sprinkle with salt, and place the item in direct sunlight to dry. Then launder as usual.

Wash Out Mildew 🏠 RATS! YOU BROUGHT your favorite leather handbag out of storage, took one whiff, and said "peee-yooo"—mildew! Don't despair. Just dampen a cotton pad with antiseptic mouthwash, and gently rub the bag's surface thoroughly. Wipe it dry with a soft cotton cloth, buff with a second cloth, then apply a commercial leather-nourishing cream. (By the way, this trick works on any mildewed leather.)

That's Brilliant!

Do you have some old lace that's gotten mildewed (say, an heirloom wedding veil, or a pair of antique curtains you picked up at a flea market)? Send that mildew packin' by rubbing the lace with a bar of mild soap until a visible film develops. Set the piece in the sun for several hours, then rinse in cold water.

Textured-Plaster-Pleasing Cleaner

> 2 tbsp. of liquid laundry detergent
> 2 tbsp. of white vinegar
> 1 qt. of warm water

If you've got textured plaster walls, you know that it's all but impossible to clean them without ripping rags, shredding sponges—and leaving bits and pieces of the stuff behind. Well, stop fretting. Mix up a batch of this simple cleaner, and (the secret to its success) grab some scraps of thick, low-pile carpet. If you don't have any remnants on hand, check with a local carpet shop for some outdated samples.

DIRECTIONS: Mix the ingredients together in a bucket. Then dip a piece of carpet into the solution and scrub the wall as usual. The carpet will glide right over the jagged plaster and get into all the nooks and crannies without leaving pieces of itself behind. When you're finished, rinse the wall using a clean piece of carpet, and blot it dry with a terry-cloth towel.

Panty Hose Particulars IT'S TOO BAD that Grandma Putt didn't live to see the debut of panty hose in 1958. In fact, she was already a grandma when nylon stockings came on the market back in 1940. Throughout the war years, they remained in such short supply that nobody could buy many of them—certainly not enough to build a stockpile. If you have some old panty hose, you can use them to scrub walls, as well as windows, furniture—and even nonstick cookware. Just wad the panty hose into a ball, and use it as you would any sponge or scrub pad.

Q *Whenever I buy new wallpaper or carpet, I always overestimate how much I'll need, and I end up with an awful lot of leftovers. Is there anything I can do with it all?*

A You can keep a scrapbook! When you put up new wallpaper, add a border to a room, or install new carpet, save some of the scraps to use for testing cleaning products and techniques. That way, you can experiment to your heart's content—without risking damage to the real thing.

Tough Treatment for Stubborn Ink Stains

1 tbsp. of milk
1 tbsp. of white vinegar
1 tsp. of borax
1 tsp. of lemon juice

Got an ink stain that just won't budge? Scrubbing alone isn't going to do anything, and it may end up driving the ink even deeper into the fabric. So when you've exhausted the old tried-and-true stain-removal methods, try this much stronger solution.

DIRECTIONS: Mix the ingredients in a bowl, then place the stained area of the clothing between two old (but clean) rags, so the solution doesn't go directly on the fabric. Dip a sponge in the liquid and gently pat the rag that's on top of the stain for about three minutes. Then remove the top rag and use a clean sponge to blot the stain with cool water. Repeat all of the steps until the stain is completely gone, then launder the garment as usual.

Ink Out 🏠 WHEN SMOOTH LEATHER gets spotted with ink stains, dip a cotton swab into some rubbing alcohol or non-oily cuticle remover, and rub it into the stains. You can also spritz the spots with an inexpensive aerosol hair spray. Wipe away whichever cleaner you use with a soft, clean cloth until the spots disappear or at least blend in with the rest of the surface.

Super Shortcuts

✂ Ink pens can cause a lot of problems. A pen can explode in your shirt or pants pocket, leaving its mark. Pen ink can seep onto your hands and can even leak onto your wood tables. When your wood furniture is the victim, soak up the excess ink immediately. Then clean the surface with a damp cloth. Water-soluble ink should disappear. If the ink's not water-soluble, the stain will likely persist. Treat a stubborn ink stain with a mixture of rottenstone (available at hardware stores) and vegetable oil. Rub the mixture into the stain, and then dry the area with a soft, clean cloth.

Vomit Stain Vanishing Spray

> ½ cup of rubbing alcohol
>
> ½ cup of white vinegar
>
> ½ cup of water

Use this solution to quickly remove most "recycled food" stains on carpet, furniture, mattresses, or clothes. It's safe for most fabrics, but test it first in an inconspicuous area to make sure it doesn't affect the dye or otherwise damage the material.

DIRECTIONS: Mix the ingredients in a handheld sprayer bottle. Lightly spritz the stained area, wait a few minutes, and blot the stain. Repeat until all traces are gone. The alcohol in the mixture not only helps remove the stain, but also makes the solution dry quickly.

Kitchen Aid 🔼 DISHWASHING LIQUID MAKES a great treatment for vomit spots because it lifts out the grease and most food stains. Use it as a laundry pretreatment on washables. Or moisten a sponge with a mixture of dishwashing liquid and a little water, and dab stains away on upholstery, carpet, or other nonwashables. Rinse the soap out of the area by blotting it with a clean, moist cloth.

Don't Let That Stain Set In 🔼 DID THAT PIZZA SLICE accidentally drip all over your lap? To stop a greasy accident from becoming a permanent stain, you need to act fast. Go into the kitchen for some baking soda, cornmeal, or cornstarch. Sprinkle it over the stain. When it dries, brush it off, and most of the oil will go along with it. Then as soon as you have a chance, launder the garment per label instructions.

GRANDMA'S OLD-TIME TIPS

To help soak up any lingering moisture and neutralize the sickly smell of vomit, Grandma Putt would sprinkle a layer of baking soda over the area after she'd blotted up as much moisture as possible. She'd let it sit for about an hour, then she'd vacuum up the residue. You can try this trick, or use 1 part borax and 2 parts cornstarch in place of the baking soda. Either method will absorb moisture and eliminate odors.

CHAPTER 8

Happy
HOLIDAYS

Need a last-minute birthday gift? Fresh ideas for trimming the tree? Or how about tips for keeping kids occupied while you get through the bustle of special occasions? No matter what your holiday challenge, I've got you covered. I've even thrown in ideas for how you can pamper yourself when the strain of entertaining takes its toll. So whether you're prepping your yard for a Fourth of July barbecue, or hosting a house full of relatives for Christmas Eve, follow my advice to make the most of your holiday celebrations.

Bathing Beauty Bath Salts

These fragrant bath salts look beautiful displayed in glass jars on a shelf in your bathroom. And soaking in a hot bath "seasoned" with these salts will make you feel oh-so-pampered during the hectic holidays. Fill a few extra jars and give them to friends as gifts, too!

2 cups of Epsom salts

1 cup of coarse salt (like kosher salt)

Food coloring

¼ tsp. of glycerin

Citrus, floral, or vanilla oil

DIRECTIONS: In a large glass or metal mixing bowl, combine the salts and stir to mix. Slowly add your choice of food coloring, drop by drop, and stir it into the salts with a metal spoon, adding more coloring until you've reached the desired hue. Stir in the glycerin and about 5 drops of essential oil. Use a measuring cup to scoop up the salts and pour them into glass jars with lids. Add gift tags with these instructions: "To use, add ½ cup of the mixture to your bathwater, ease into the tub, and relax."

Basic Bath Salts 🏠 FOR A LESS FANCY VERSION of the Bathing Beauty Bath Salts recipe (above), mix together 1 cup each of Epsom salts, table salt, and baking soda. Store in an airtight container at room temperature. To use, add about 2 tablespoons of the mixture to your bathwater.

PINCHING PENNIES

Many tonics call for the use of extracts and essential oils, and many of these oils come in fancy-shaped, colorful bottles that are practically works of art. So don't even think about throwing them away! Get every ounce of your money's worth and turn the bottles into bud vases, or use them to hold gifts of homemade lotions, potions, or herbal vinegars. You can even reinvent them as fancy dishwashing liquid bottles. Just pop in a stopper with a pour spout, which you can usually find in housewares stores or the supermarket's kitchen-gadget section.

Bathtub Cookies

These bath-time cookies are the perfect gift for a youngster who's going through the anti-bath stage. They also make great gifts for grown-ups who love nothing better than a long soak in the tub.

2 cups of fine-grain sea salt*
½ cup of baking soda
½ cup of cornstarch
2 tbsp. of light vegetable oil
1 tsp. of vitamin E oil
6 drops of your favorite scented oil
6 drops of food coloring (optional)

DIRECTIONS: Mix all of the ingredients together, roll out the dough, and cut out shapes with cookie cutters. Bake at 350°F for 10 to 12 minutes. (Don't overbake!) Let the cookies cool completely, and put them in a big glass jar or other decorative container. To use, just add one or two cookies to the bathwater, and enjoy!

*If your supermarket doesn't have fine-grain sea salt, use the coarse type instead, and run it through your blender or coffee grinder.

Fancy Soap Holders 🏠 REMEMBER ALL THOSE big scallop shells or clamshells you collected on your last beach vacation? Turns out they're just the ticket for packaging your gifts of homemade soap (see Fun and Fancy Soap, page 205). First, spiff up the shells with water and an old toothbrush. Then, once they're dry, place a bar of soap in each shell and wrap the whole shebang in cellophane or tissue paper.

That's Brilliant!

Let the lucky recipients of your homemade soap (see Grandma Putt's Soap Balls, page 207) know that you made it just for them by adding a personal touch to the bars. Use your sister's favorite color to dye her soap, or cut the bars in the shape of a flower to personalize your gardening grandma's soap. You can even customize the soaps by adding different exfoliants, like oatmeal, coffee grounds, or sugar. There are plenty of ways to personalize the soaps—and the gifts will be appreciated that much more.

Bombastic Bubble Solution

> **2 parts dishwashing liquid**
> **2 parts water**
> **1 part vegetable oil**

For kids, every day feels like a holiday when they've got bubbles! And they don't need to wait for anyone to bring them a store-bought solution. You can help them stir up a never-ending supply right at home using this simple recipe. Then let 'em rip!

DIRECTIONS: Mix the ingredients together in a shallow bowl or tub. Then have the kids dip their bubble blowers into the fabulous fluid, and make lots and lots of marvelous, miraculous bubbles.

Super Bubble Solution 🏠 IN THE SUMMER, the kids in my neighborhood used to make giant bubble blowers by bending a wire clothes hanger into a circle (more or less), leaving the hook in place to use as a handle. Then we'd pour some Bombastic Bubble Solution (above) into a shallow pan, dip the loop into the "drink," and draw it gently through the air—making the biggest bubbles I'd ever seen.

A Tisket, A Tasket 🏠 YOU CAN USE those plastic berry baskets for more than just carrying berries. Mix up a large bucket of dishwashing liquid and water, then dip the basket in. Wave it around in the air—it makes an amazing blast of bubbles!

A Sparkling Send-Off 🏠 BLOWING BUBBLES at newlyweds has become ordinary. So if you're planning a summer wedding (especially around the Fourth of July), be extraordinary and ask your guests to light sparklers to mark the occasion. Trust me, the photos will be absolutely spectacular.

PINCHING PENNIES

Don't waste your money on store-bought bubble bath to give as a gift to a young friend. Instead, whip up your own sudsy solution by mixing ½ cup of dishwashing liquid and 1 tablespoon of vegetable oil in ½ cup of water. Then pour the solution into a brightly colored squeeze bottle and place it in a gift bag along with a kid-friendly washcloth and towel set. The lucky recipient will have tons of fun playing in the bubbles—and he or she will get clean at the same time (so it's a gift for Mom and Dad, too!).

Easy Herbal Vinegar

> ½ cup of dried herbs*
> 2 cups of white vinegar

Just about anything that plain vinegar can do, herbal vinegar can do better! Besides adding pizzazz to any recipe, it's like a combination first-aid kit and beauty spa in a bottle, so it makes the perfect gift.

DIRECTIONS: Put the herbs in a sterilized glass jar, and mash them slightly with a spoon. Pour the vinegar over the herbs, cover the jar tightly, and let the mixture steep in a dark place at room temperature for one to three weeks. Shake the jar every few days, and taste the vinegar once a week until the flavor reaches the desired intensity. Add more herbs if you need to. When the flavor suits you, strain out the solids, and pour the vinegar into a sterilized decorative bottle with a tight cap. Be sure to tell the recipient to keep the vinegar away from direct light, which rapidly decreases the herbs' potency.

*Or use 1 cup of loosely packed fresh herb leaves, washed and dried.

Choosing Your Herbs 🏠 WHAT KIND OF HERBS should you use for your Easy Herbal Vinegar (above)? It all depends on what you want them to do for you. Here's a sampling of common supermarket options:

   Bay leaves: antiseptic, freshen all skin types; **marjoram and oregano:** antiseptic, soothe sore throats, relieve aching muscles; **mint:** relieves headaches, cools and refreshes; **parsley:** cleanses oily skin, lightens freckles, adds shine to dark hair; **rosemary:** antiseptic, repels insects, good for oily skin; **thyme:** antiseptic, deodorizes.

GRANDMA'S OLD-TIME TIPS

Every year, Grandma Putt would get a fresh pine tree for Christmas. It was always beautiful, but those pine needles were messy. Yet Grandma—who found a clever use for everything—put them to work as potpourri instead of pitching them. She'd pile the needles in attractive bowls, and when they dried out, she'd toss them into the fireplace.

Egg-stra-Special Easter Egg Dye

Eggs
Water
1 tsp. of white vinegar
2½ cups of plant dye
    material

Want to try your hand at an all-natural way to dye Easter eggs? It's easy! All you need are a few colorful purchases from the produce department. Start with these color options, then experiment with your own: blueberries (for blue); grape juice or crushed violet blossoms (for purple); instant coffee crystals (for brown); pomegranate juice, red onion skins, or raspberries (for red); shredded beets or cranberries (for pink); shredded carrots or lemon peels (for yellow); spinach (for green).

DIRECTIONS: Place the desired number of eggs in a saucepan, and cover them with water. Add the vinegar, then the coloring ingredient of your choice. Boil for 30 minutes, remove the pan from the heat, and check the color of the eggs. For more intense hues, continue soaking the eggs in the colored water for up to 12 hours.

Fancy Eggs ⬆ WANT TO FILL the kids' Easter baskets with the fanciest eggs in town? All you need are some old panty hose, food coloring, and an assortment of tiny leaves or flowers—and hard-boiled eggs, of course! First, make your egg dye following the directions on the food-coloring package. Press a leaf or flower flat against an egg, wrap a piece of panty hose around it, and tie it tightly with string. Dip the egg into the dye, keep it there until it reaches the color intensity you want, and then pull it out. Let the egg dry inside the panty hose, cut away the fabric, and voilà!—a one-of-a-kind egg designed by you!

That's Brilliant!

Here's a simple way to make decorative patterned eggs for the kids' Easter baskets. First, lay a paper towel over a piece of aluminum foil. Then dribble 8 to 10 drops of food coloring on the towel. Set a damp, hard-cooked egg on top, and gently wrap the paper and foil around the egg. Wait a few seconds, then open the "package" and let the egg dry.

Every Day Is Christmas Potpourri

Love the holidays, but hate when it's time to take down the decorations and haul the tree to the curb? Our house smelled like Christmas all winter long even without the tree, thanks to an extra-simple potpourri that Grandma made every year.

DIRECTIONS: Combine the orrisroot and pine oil, then pour the mixture over the pine needles until they are evenly coated. Place the piney potpourri in decorative bowls and baskets, and set them wherever you'd like to spread an aromatic uplift.

*Available in craft-supply stores, herb shops, and nurseries that specialize in herbs.

Blanket Your Beds with Branches

After the holidays are over, saw the branches off your Christmas tree and lay them across your sleeping perennial beds. Arching them over the plants will protect the sleeping beauties from snow, frost, and damaging winds. And you can easily check on them whenever you want by lifting the branches, then gently putting them back in place until springtime rolls around.

Artificial Branches

IF YOU HAVE an artificial Christmas tree, you know how hard it can be to get the thing apart after the holidays are over. End that yearly struggle by dipping the end of each branch into petroleum jelly before you insert it into the tree trunk. When you're ready to "undecorate," it'll slide right out.

PINCHING PENNIES

What can you do with all those CDs and DVDs that keep arriving in the mail, pitching everything from Internet services to new supermarkets? Use them to make a shimmering techno-tree at Christmastime. Drill a hole in the rim of each disc, and decorate one or both sides with paint, sequins, fabric scraps, or other pretty stuff. When you've got a whole collection of ornaments, tie ribbons through the holes, and hang 'em up.

Extra-Easy Modeling Clay

Kids love playing with modeling clay. It's a great, fun activity for any day of the year. So mix up a batch of this easy-to-make, nontoxic clay for gift giving. It's safe enough for small children to use, but long-lasting enough to please even serious crafters.

> 1 part all-purpose flour
> 1 part cornstarch
> 1 part white glue
> Food coloring (optional)

DIRECTIONS: Combine all of the ingredients in a bowl, then turn the mixture out onto a bread board, and knead the lump until it's the consistency of bread dough. Add more flour and cornstarch, or more glue, if you need to. Package the clay in a sealed container, with a gift note saying that creations like sculptures, jewelry, or Christmas tree ornaments can be air-dried and decorated with acrylic paint.

Clay Your Way ⬠ MAKE MODELING CLAY this way: In a saucepan, mix 2 cups of flour, 1 cup of salt, 4 tablespoons of cream of tartar, 2 tablespoons of vegetable oil, a few drops of food coloring, and 2 cups of water. Stir over medium heat for three to five minutes, or until the mixture forms a ball. When it's cooled, knead it with your hands, and store it in an airtight container.

Bathtub Fun ⬠ WHEN YOUNGSTERS are visiting for the holidays, offer them this fast, easy, and inexpensive treat—just add a few drops of food coloring to their bathwater. The kids will have a blast at bath time, and will probably even beg to get all washed up every night!

That's Brilliant!

Make a kid-safe (and furniture-safe) paste by mixing flour and water to a pancake-batter consistency. Then, you'll be able to give your kids or grandkids a paste they can use on any medium. This adhesive works like a charm on paper, lightweight fabric, and cardboard.

Fake Snow Farewell Formula

Snow in a can is a riot to use at Christmastime to give your windows that frosty look. Find the spray in the Christmas decoration aisle of any discount store, and spray to your heart's content. Then, when Christmas is over, use this easy trick to melt away that fake snow—in less than a minute!

½ cup of rubbing alcohol
⅛ cup of ammonia
A few drops of dishwashing liquid
3 cups of water

DIRECTIONS: First, brush off all of the loose bits of fake snow with a dry cloth, and then vacuum up any stragglers. Mix all of the ingredients in a handheld sprayer bottle, and thoroughly spray the snow. Wait 30 seconds or so, and wipe it away. Your window will dry snow- and streak-free!

Coffee Filter Snowflakes 🏠 IF YOU BREW YOUR morning joe at home, you've probably got a stack of coffee filters on hand. And, since they're so cheap, you can spare a few for a fun project around the holidays. Just fold a filter in eighths, cut out shapes along the edges, then open it up to reveal your snowflake creation. When you've made a full set of flurries, tape them onto a window, or hang them from the Christmas tree.

GRANDMA'S OLD-TIME TIPS

Even though Grandma never had access to cooking oil in a spray can, she still had plenty of oil around when she was cooking. And she had 101 different uses for it, too. One of her favorites (but definitely not one of Grandpa's) was to lighten the load on the snow shovel when three or four inches were dropped on us. She would just coat the shovel with cooking oil, and let Grandpa go out and do his job. The white stuff slid right off—making the job faster and easier (so Grandpa had more time for other chores).

Fun and Fancy Soap

Looking for a creative school holiday project that can keep the kids entertained for hours (and produce a batch of fun gifts)? Then let them get a little soapy. Kids and grown-ups both love this colorful, easy-to-make soap.

1 bar of Ivory® soap, grated
4 drops of food coloring
2 to 3 drops of orange, peppermint, or vanilla extract
¼ cup of warm water

DIRECTIONS: Combine all of the ingredients in a bowl, and stir until the mixture starts to stiffen. Remove it from the bowl, and knead it until it reaches the consistency of very thick dough. Spoon the mixture into plastic molds or cookie cutters (make sure the top of the dough is smooth), and set them in the freezer for 10 minutes. Then pop the soaps out of the molds, and let them dry until they're hard. For gift giving, wrap individual soaps in tissue paper and tie with a ribbon.

Those Yellow Lemon Thingies 🏠 I KNOW—PRETTY SCIENTIFIC. But I don't know what else to call those little plastic lemons that typically hold lemon juice. I do know that they can double as soap dispensers, which is what my wife uses them for. I thought it was a little strange at first, but it works and she's happy, which, as you guys out there know, means that I'm happy too! By the way, they can also double as gift packaging for your homemade liquid soaps.

Super Shortcuts

✂ Looking for another great activity to keep kids busy as their school holiday break drags on? Use some food coloring to make finger paints for your junior artists. Mix ¼ cup of cornstarch with 2 cups of cold water, and boil, stirring constantly, until the mixture thickens. Pour it into small containers, and add a different food coloring to each one. Then bring out the scrap paper, and watch your budding artists go to town.

Funny Bunny Sidewalk Chalk

Give the kids a colorful surprise from Good Ol' Peter Cottontail with a plastic egg full of chalk! It's fun to make and sturdy enough to decorate all the sidewalks in the neighborhood.

DIRECTIONS: Pull apart the halves of the plastic eggs, then generously coat the insides of each half with petroleum jelly. Set each egg half in the egg carton, open side up. For each color, use a disposable spoon and a paper cup to mix together ¼ cup of plaster of Paris, 2 tablespoons of water, and 2 tablespoons of powdered paint. Spoon the mixture into a top and a bottom of an egg, filling them almost to the top. Let the mixture sit until it has a thick, muddy consistency, then snap the egg together and shake it briskly to allow the contents to mix. Repeat the steps, starting with a fresh plastic spoon and paper cup for each color. Let the chalk eggs sit in the carton overnight, and then remove them from the molds, using a knife to help loosen the pieces if necessary.

*Available at craft-supply stores; tempera paint is also referred to as poster paint.

> 6 plastic eggs
> Petroleum jelly
> 1 empty egg carton
> 1½ cups of plaster of Paris
> 2 tbsp. each of 6 different colors of powdered tempera paint*
> ¾ cup of cold water

Eye-Popping Easter Eggs 🥚 WHEN EASTER ROLLS AROUND, make mini works of art to tuck into your baskets. Just gather up some rubber bands of various sizes, and wrap them around each egg before you dip it into the dye. The finished product will come out with stripes or diamond patterns. And each will be a one-of-a-kind creation, because you'll never be able to get the bands in exactly the same spots on every egg.

That's Brilliant!

You used to buy panty hose that came packaged in big plastic eggs, and you kept every single one of them because you knew they'd come in handy one day. Well, that day is here! Haul those things out of the attic, and decorate them with paint, paper, fabric, or spangles. Then fill them with jelly beans, marshmallow bunnies, and tiny treasures, and tuck them into Easter baskets.

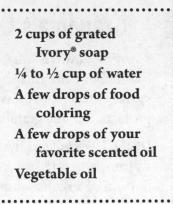

Grandma Putt's Soap Balls

My Grandma Putt always had balls of soap in her guest bath, and I loved using them because of their smooth round shape. This old-fashioned recipe for soapy dough can be easily shaped into balls that make a thoughtful birthday or Mother's Day gift. And they're so easy to whip up, even kids can get into the act.

2 cups of grated
    Ivory® soap
¼ to ½ cup of water
A few drops of food
    coloring
A few drops of your
    favorite scented oil
Vegetable oil

DIRECTIONS: Mix the grated soap with the water by hand to form a smooth paste. For lighter, fluffier soap dough, use an electric mixer. Add more water or soap as needed until the mixture is smooth. Stir in the food coloring, then blend in a few drops of scented oil and mix well. Lightly coat your hands with some vegetable oil and form the mixture into small balls. Place the soap balls on a sheet of wax paper and let them harden for several hours. To give as gifts, wrap the balls individually in colored tissue paper, or leave them unwrapped and put them into a decorative bowl, tin, or basket.

Trash to Treasure 🏠 DON'T THROW AWAY those fabric scraps, folks! If you're giving homemade soap as a gift, use pinking shears to cut the edges of a pretty square of fabric, set the bar or ball of soap in the middle, pull the corners up, and tie the whole shootin' match with a beautiful ribbon.

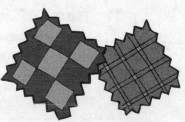

When your bars of soap dwindle down to slivers, don't pitch 'em! Instead, stuff them into a panty hose leg. Then tie a knot in the hose above the soap, and use twine to attach a card with these instructions: "Tie this trash-to-treasure soap bag to your outdoor spigot for handy hand cleanups!" Stick the soap sack into a basket that includes garden gloves, seed packets, and other miscellaneous outdoor supplies, and give the gift to your favorite gardener.

Grandma's Crystal Palace

> ½ cup of water
> ¼ cup of Epsom salts

When I was a lad, Grandma Putt always ushered in Christmas by helping me make these sparkly treasures. It was one of my favorite holiday projects, and my grandchildren still get a kick out of growing their own crystals.

DIRECTIONS: Boil the water in a pan. Remove it from the heat, pour in the Epsom salts, and stir until they're completely dissolved. Put a sponge in the bottom of a shallow bowl, and pour the Epsom salts solution over it. Put the bowl in a sunny spot, and keep a close watch on it. As the water evaporates, crystals will form all over the sponge, creating what looks like a miniature ice palace. Best of all, no two crystals are alike, so you'll get a new result each time!

Sponge Animals 🏠 WANT TO MAKE a handful of stocking stuffers that'll make bath time more fun for the youngsters in your life? Buy a pack of sponges from the grocery store and use sharp scissors to cut them into kid-friendly shapes, like boats, flowers, or favorite animals. Then cut a slit in one end of each sponge and tuck in a tiny, travel-size bar of soap. Bingo—just add water for a super-soaping tub toy!

The Name of the Game 🏠 ADD A PERSONAL TOUCH to your gift of sponge animals (see above) by cutting the letters of the recipient's name out of small sheets of craft foam. The young bather can get the letters wet in the tub and then stick them to the tile wall to practice spelling his or her name.

Super Shortcuts

In my younger days, I learned how to grow an indoor "lawn," and it quickly became one of my favorite wintertime projects. Here's all there is to it: Find a natural sponge with large pores (not the synthetic kind), and soak it in water for a couple of minutes, until it's sopping wet. Squeeze it until it's about half dry, and sprinkle grass seed into the openings on top. Then set the sponge, grass seed side up, on a plate in a sunny window, and sprinkle it lightly with water every day. Before you know it, you'll have a tiny patch of green to remind you that spring is just around the corner!

Homemade Finger Paint

Need something to distract the kids or grandkids while you wrap presents? Just whip up a batch of finger paint—the old childhood standby that's adored by generations of kids. It's a great activity that can lead to an entire afternoon of fun. Just make sure the little rascals keep the paint on the paper!

DIRECTIONS: Stir together the gelatin and ¼ cup of the cold water in a large bowl. In a saucepan, combine the cornstarch and the rest of the cold water until the mixture is smooth. Add the hot water and cook over medium heat, stirring constantly, until the mixture boils. Remove it from the heat, and add the gelatin mixture. Stir in the soap flakes until they dissolve. Pour a little bit of the mixture into different containers, and add ½ tablespoon of tempera paint to each container. Now it's time to paint a masterpiece!

*Available at craft-supply stores; tempera paint is also referred to as poster paint.

- 1 envelope of unflavored gelatin
- 1 cup of cold water
- ¾ cup of cornstarch
- 2 cups of hot water
- ½ cup of mild laundry soap flakes
- Assorted colors of liquid tempera paint*

Keep It Clean YOU DON'T HAVE to use the stove to mix up your Homemade Finger Paint (above). You can cook the mixture in the microwave until it comes to a boil and is nice and smooth. Also, for a more kid-friendly option, you can substitute food coloring for the tempera paint.

That's Brilliant!

Have your young artists run out of finger paint? Make a new supply by squirting shaving cream into small bowls and adding a few drops of food coloring to each one (the more you use, of course, the darker the shade will be).

Kid-Pleasing Play Clay

There's no greater activity for children than sinking their hands into mushy modeling clay. It's a wonderful way to get creative without making too big of a mess—because cleanup's a cinch! Your young sculptors can turn out Christmas tree decorations galore with this kitchen-counter modeling medium.

> 2 cups of baking soda
> 1 cup of cornstarch
> 1½ cups of water
> Food coloring (optional)

DIRECTIONS: Mix the baking soda, cornstarch, and water together in a pan, and cook over medium heat, stirring constantly, until thickened. Spread the mixture on a plate or cutting board, cover it with a damp cloth, and let it sit until it's cool enough to handle. Knead the dough until it's smooth, adding food coloring if you'd like. (Use as much as it takes to reach the shade you want.) Store the clay in an airtight container in the refrigerator.

Drying Play Clay ⌂ YOU HAVE TWO drying options for the finished play clay creations: Let them sit, uncovered, for a few days, or bake them in the oven on the lowest setting for half an hour or so, checking every few minutes to make sure they don't "overcook." (With either method, drying time will vary, depending on the thickness of the objects.) When they're good and dry, the artist can hang them on the tree unadorned, or color them with acrylic paint.

Q My kids got crayon all over their clothes on Christmas Eve, just before we were leaving for church. When I washed their clothes later, the stains hadn't budged. How can I get the marks out?

A If your young Picasso colored his clothes as well as his paper, banish crayon stains by adding a box of baking soda to the wash cycle, along with your regular laundry detergent.

Lemony Rosemary Soap

The scent of citrus seems to make everything feel a little more clean and fresh. Add to that the scent of rosemary, and you've got a combination that can't be beat. The lemon juice and rosemary in this soft, creamy soap recipe will refresh and tone all skin types, so it's an ideal gift.

¼ cup of finely grated castile soap

¼ cup of lemon juice

2 tbsp. of almond oil

2 tbsp. of cream of tartar

1 or 2 tbsp. of finely chopped rosemary leaves

DIRECTIONS: Place the castile soap in a double boiler set over simmering water, and stir until melted. In a small bowl, mix together the melted soap with the rest of the ingredients until it's the consistency of honey. Pour the mixture into a ceramic ramekin and cover with colorful plastic wrap, held with a decorative band. Tell any recipients of this gift to scoop out a small amount with their fingers whenever they need to refresh or clean their hands or body.

A Liquid Asset 🏠 DISHWASHING LIQUID BOTTLES are handy gadgets to keep around and give away. Once the soap's gone, wash out the bottle and use it for everything from watering plants and dispensing cooking oil to squeezing out pancake batter in funky shapes. Plus, as my grandkids will tell you, they make awesome squirt guns! Just make sure you cover them with Con-Tact® paper to suit the occasion before you give them as gifts.

Color Your Glue 🏠 IF YOU LIKE to make your own holiday greeting cards, try this technique for a creative change: Fill a squeeze bottle with white water-based glue, then tint the glue with a few drops of food coloring. You (or your kids) can use it to make raised designs on paper.

PINCHING PENNIES

Got an empty berry basket? Give it a spin through the dishwasher, and when it's dry, decorate it to your heart's content. You can cover it with papier-mâché and paint it. Or you can weave ribbons and bits of raffia through the plastic mesh. Then fill your now-beautiful basket with seasonal goodies and give it as a hostess gift.

Perfectly Pleasing Pomanders

What the heck is a pomander?! It's a piece of fruit, studded with cloves and rolled in spices. Once you let it dry, it's the perfect way to add fragrance to your home around the holidays.

DIRECTIONS: Run 1 strip of tape (sticky side out) around the fruit vertically, then run a second strip horizontally, dividing the fruit into quarters. Pin the tape into place where needed to keep it from falling away. Pierce a small section of an untaped part of the fruit with the pointed ends of the cloves, as close together as possible. Remove the cloves, and use the piercing tool to make similar-sized holes in the rest of the untaped portions. Put cloves in all the holes. Mix the orrisroot and pumpkin pie spice in a bowl, and roll the fruit in it several times a day for two or three days. The spice will stick to the fruit where the juices have seeped out. Let the fruit sit in a dry area for two to five weeks. Once it's dry, remove the tape and add some decorative ribbon in its place.

*Available in craft-supply stores, herb shops, and nurseries that specialize in herbs.

1 piece of firm, unblemished fruit (apple, lemon, lime, orange, or pear)

½-inch-wide adhesive tape

Straight pins

2 oz. of whole cloves

Piercing tool, such as an ice pick, awl, or knitting needle

¼ cup of powdered orrisroot*

¼ cup of pumpkin pie spice

Pomanders for Any Season POMANDERS DON'T HAVE to be just a Christmastime gift. In the spring, exchange the spice mixture for floral potpourri. It'll add a nice texture and springtime scent to the pomander.

Super Shortcuts

If you're having trouble handling your pomander fruit while you insert the cloves, here's a trick to make it easier. Secure the flat head of a 4-inch nail into some modeling clay, then pierce your pomander with the pointed end so that the fruit sits on top of the nail. That way, it won't slip while you push in the cloves. For added help, wear a thimble on your thumb so it won't get sore from all the pushing.

Salty Sculpture Clay

When your children or grandchildren have time on their hands at a school holiday break, help them get their creative juices flowing by keeping them well stocked with modeling clay. This easy-to-make formula works just as well as the commercial brands, and it's a whole lot cheaper.

> 1 cup of all-purpose flour
> 1 cup of salt
> 1½ cups of warm water
> 2 tbsp. of vegetable oil*

DIRECTIONS: Combine the flour and salt in a medium-size bowl. Add the water, then mix until a dough forms, and roll it out as you would cookie dough. If the dough gets too sticky to work with, put in a little more flour.

*Add vegetable oil to the mixture only if you intend to store the clay for use at a later time.

Salty Snow Sparkle TRANSFORM YOUR HOME into a winter wonderland by turning ordinary candles into glistening decorations. It's easy! Just paint a candle with a light coat of glue, then roll the candle across a layer of Epsom salts that you've poured onto a sheet of aluminum foil. Shake off the excess and allow the candle to dry upright for at least an hour before handling it.

More Fun with Salts TO FROST YOUR WINDOWS, pour ¼ cup of Epsom salts into 1 cup of beer (avoid dark-hued brews), and allow the mixture to sit for at least 30 minutes until the salt has mostly dissolved and the foam has gone away. Use a sponge to paint the mixture onto windowpanes, dabbing random sections with a paper towel. Allow the windows to dry overnight, and in the morning, you'll wake up to windows that are so icy looking, Jack Frost will wonder who's taken over his job! When the holidays are over, simply wipe the frost away with window cleaner.

That's Brilliant!

Did Santa leave a brand-spankin' new pair of blue jeans under your Christmas tree? If you want your jeans to stay a dark, rich blue but lose that stiff-as-a-board texture, here's a tip for you: When you wash a pair of jeans for the first time, add ¼ cup of salt to the wash cycle. It will help lock in the color and soften the fabric.

Scented Christmas Tree Ornaments

1 cup of all-purpose flour
1 cup of pine-scented*
   potpourri
½ cup of salt
Food coloring (optional)
⅓ to ½ cup of water
Cookie cutters**
Yarn or ribbon

Although many kinds of evergreen conifers make great-looking Christmas trees, some of them have about as much aroma as a paper bag. To avoid disappointment, tuck this recipe into a box with your Christmas decorations. Then, if you bring home a tree that's low on fragrance—or if you opt for the artificial version—you'll be all set to whip up a few batches of these ornaments. They'll give any tree that old-time scent that says, "Santa Claus is comin' to town!"

DIRECTIONS: Combine the flour, potpourri, salt, and food coloring (if desired) in a bowl, and stir in the water until the mixture is the consistency of cookie dough. Put it in the refrigerator for five minutes. Using a lightly floured rolling pin, roll out the dough to a thickness of about ¼ inch, and cut it with cookie cutters. Poke a hole in each ornament, and set them on a wire rack to dry overnight. Then hang them on the tree with yarn or ribbon.

*Or substitute another fragrance of your choice.
**Use open cookie cutters, not the kind with closed tops.

Ornament Repair 🔨 A FEW SCRATCHES on a glass Christmas tree ornament can add character. But if the finish is peeling off in big patches, that's another story. You can still save those decorations by stripping off the old paint with a half-and-half solution of ammonia and water, then rinsing with clear water. When the baubles are completely dry, paint them with glossy enamel (brush or spray it on), and hang them on the tree.

Super Shortcuts

✂ Before you hang a Christmas wreath outside, spray the ribbons and bows with super-hold hair spray and let it dry. Now they'll stay clean and perky. (Just hang the wreath in a sheltered spot where it won't get rained on, or the hair spray will wash right off.)

Scent-Sational Holiday Elixir

Keep this terrific tonic simmering on your stove for a holiday treat for your nose! You can mix the dry ingredients ahead of time and store them in a ziplock plastic bag. Then follow the instructions to heat and release their aroma.

4 cinnamon sticks
2 bay leaves
⅓ cup of whole cloves
¼ cup of whole allspice
2 lemon slices
2 tbsp. of orange peel
½ tsp. of ground nutmeg
1 qt. of water

DIRECTIONS: Mix all of the ingredients in a medium saucepan and simmer on low heat. Add more water as the liquid evaporates, and your house will smell like a holiday wonderland all day long.

Christmas Ornaments 🏠
IF YOU'RE FEELING CREATIVE, use burned-out lightbulbs to make Christmas ornaments. You can paint them, decorate them with ribbons, make faces, or whatever strikes your fancy. By the way, I recently saw some beautiful lightbulb ornaments in a very upscale home decor shop—so if you're really good at this, you could end up not only saving money, but making some extra moolah, too!

Needle Picker-Upper 🏠
WHEN THE TIME COMES to untrim our Christmas tree, we always spread a sheet around the base. That way, any loose needles fall onto the fabric. After we remove the ornaments, we just wrap the sheet around the tree, haul it outside, and deck it out with edible treats for the neighborhood birds.

PINCHING PENNIES

Okay, now I know that every one of you out there has at least a dozen empty wire coat hangers cluttering your closets. After all, it's common knowledge that if you put a coat hanger in a closet and shut the door, by morning, there'll be two more in there! But, you can decrease your inventory by reinventing them. Bend them into seasonal shapes for Christmas, Halloween, and other holidays, then hang 'em on the walls or from light fixtures for some fun, funky design accents.

Super-Simple Lavender Soap

> 10 tbsp. of finely grated castile soap
>
> 8 tbsp. of boiling water
>
> 2 tbsp. of dried lavender flowers, crushed into a powder
>
> 4 drops of lavender oil

It seems like everywhere you look these days there's a bottle of hand sanitizer standing by. And, of course, it's important to keep your hands clean so germs and other toxins don't have a chance to get you sick. But the smell of the stuff . . . ugh! So bypass the commercial concoctions and whip up a batch of this lavender soap, which makes hands squeaky clean—and sweet smelling, too!

DIRECTIONS: Put the soap and boiling water in a bowl, then set the bowl over a pot of hot water. Stir until the soap and water mixture is smooth, then remove the bowl from the hot water. Add the powdered lavender flowers and lavender oil, then stir to mix thoroughly. Pour the soap into a decorative bottle for gift giving, or a simple plastic container if you're keeping it all to yourself.

Refresh Your Skin 🏠 WHEN YOU'RE SPENDING a scorching day under the sun at a Fourth of July or Labor Day picnic, keep your cool by misting your skin with this elixir: Mix 2 teaspoons of witch hazel, 12 drops of lavender oil, and 10 drops of peppermint oil in an 8-ounce spray bottle, filling the balance of the bottle with water. Keep it on ice, and reach for it anytime you feel too darn hot.

GRANDMA'S OLD-TIME TIPS

When Grandma Putt got stressed out over the hustle and bustle of holiday preparations, she would add a few drops of lavender oil to her favorite body lotion. That turned her lotion into a super source of anxiety-preventing aromatherapy that lasted all day long. For a more dramatic effect, add a drop of lavender oil to a dab of your favorite salve and rub it into your temples or abdomen as a reminder to relax and breathe deeply.

Sweet and Saucy Christmas Ornaments

You can give the entire house (or at least the area where the Christmas tree stands tall) a sweet holiday scent by using cinnamon. Your holiday guests will think that you're baking up something sweet the minute they take a good whiff.

> 1 cup of ground cinnamon or cinnamon blended with ground cloves, ginger, or nutmeg
>
> 1 cup of applesauce, room temperature

DIRECTIONS: Mix the cinnamon (or cinnamon-spice blend) and the applesauce in a bowl. Make sure the consistency is doughy and smooth. Spread the dough out on wax paper, and sprinkle it with some more cinnamon. Use holiday cookie cutters to cut shapes into the dough, and poke holes near the top large enough to slip a ribbon or ornament hook through. Put the wax paper and dough aside, and let the ornaments air-dry for three or four days. Turn them over a few times each day to prevent sticking. Soon, you'll have ornaments that you can put on the tree. But don't let kids or pets eat 'em—consuming such a large amount of cinnamon can be harmful to your health. **Note:** The ornaments will shrink as they dry, so make sure the holes you poke aren't too small or too close to the edges.

Consistency Is the Key 🏠 IF YOU'RE TURNING to homemade ornaments for your favorite holidays, go for consistency throughout the entire presentation. For example, if you're making candy cane–shaped ornaments, use red and white ribbons to hang them. Or, if you're giving your house an Irish twist for St. Patrick's Day, put only green ribbon through your shamrock shapes.

PINCHING PENNIES

Umbrella lights that attach to the ribs of patio umbrellas provide attractive patio lighting. Problem is, they sure are pricey! So save yourself a bundle by purchasing mini Christmas tree lights during the after-Christmas sales. Or just don't put your normal Christmas lights too far out of reach once the holidays are over. String them under the umbrella for that high-end copycat look at a fraction of the price.

Tub-Time Toddy for the Body

2 cups of milk
1 cup of honey
1 cup of salt
¼ cup of baking soda
½ cup of baby oil

Getting the yard and garden in shape for outdoor Memorial Day and other summer holiday celebrations can really wear a guy out. After a long, hard day in the yard—or even a short, easy one—nothing feels better than sinking into a tub full of hot water and this terrific toddy for a tired body.

DIRECTIONS: Combine the milk, honey, salt, and baking soda in a large bowl. Fill your bathtub with water, pour in the mixture, and then add the baby oil (and, if you'd like, a few drops of your favorite fragrance).

Soak with a Sack 🎁 IF YOU'RE THINKING ABOUT giving the Tub-Time Toddy for the Body (above) as a hostess gift, great idea! Just layer the ingredients in a glass jar with a lid. Then place the jar inside a muslin bag, and wrap it with a ribbon. Attach a card with these instructions: "Pour the contents of the jar into the muslin bag. Let the water run through the bag as the tub is filling, then use the bag as a washcloth to massage the skin." Your lucky recipient will enjoy a relaxing, invigorating soak!

That's Brilliant!

There's nothing like a good hot bath after a busy day on your feet to soothe tired muscles. After all the time you've spent making your holiday guests feel special, you deserve some pampering. A hot bath will relax you, but try adding some spice to your tub time! Grate ¼ cup of fresh ginger, and add it to the water while the tub is filling. Not only does this make the water (and you) smell great, but the ginger also promotes perspiration, which is your skin's way of getting rid of some of the toxins your body has absorbed.

CHAPTER 9

PET
Projects

Pets are part of the family. We play with them, coddle them, and turn to them for companionship and love. So it's only natural that we want what's best for our pets. When they're sick, injured, or just unhappy, we're unhappy, too. That's why I've gathered dozens of tips and tricks to help you take care of your furry friends. Read on for advice on fighting fleas, learn grooming how-to, follow my simple recipes for yummy treats, and much more. Fido and Fluffy will be glad you did!

Anti-Digging Formula

Although some breeds of dogs are more prone to digging than others, the urge is part of their basic nature. If you've got a dog who loves to excavate where he's not supposed to, or neighborhood pups are making a holey mess of your yard, reach for this spicy solution.

DIRECTIONS: Put all of the ingredients into a bucket and let the potion stand overnight (preferably in the garage or shed, where you won't smell it!). The next day, strain out the solids and pour the liquid into a handheld sprayer bottle. Then thoroughly spritz the areas where dogs are digging—or any other places you'd like to keep off-limits to canines or any other critters. **Note:** You will have to respray every few days, as well as after every rain or every time you water the area.

Get Outta My Hair! ♠ THE NEXT TIME YOUR DOG or cat has a run-in with a bush and winds up with fur matted with burrs and prickles, reach for some oil. Pour a few drops of vegetable or mineral oil on the clinging hitchhikers, then gently comb your pet. The stickums will comb right out and your pet will be happy as a clam.

That's Brilliant!

There are some simple tips that you can live by to keep your pets happy and flea-free all summer long. For one thing, make sure your pet is getting a good diet and plenty of exercise. Tests show that critters in tip-top shape attract far fewer fleas than their less healthy counterparts. Also, vacuum, vacuum, vacuum. Unless you keep fleas out of the house, you can't keep them off your pets. And give your dog or cat a dose of garlic and/or brewer's yeast every day to repel fleas. (Some folks say this is an old wives' tale, but it's kept my old pal flea-free for years.)

Beat It, Dogs! Elixir

Whether it's a border collie rolling in your begonias or a dachshund digging in your daisies, the pitter-patter of pet paws can do a real number on your flower beds. Either way, you end up with smashed plants, compacted soil, and lots of frustration. What's the solution? Try this terrific tonic to let Rover know he's not welcome there.

> 2 garlic cloves
> 2 small onions
> 1 jalapeño pepper
> 1 tbsp. of cayenne pepper
> 1 tbsp. of chili powder
> 1 tbsp. of dishwashing liquid
> 1 tbsp. of hot sauce
> 1 qt. of warm water

DIRECTIONS: Finely chop the garlic, onions, and jalapeño pepper, and then combine with the rest of the ingredients. Let the mixture sit and marinate for 24 hours, strain it through cheesecloth or old panty hose, then sprinkle the resulting liquid on any areas where dogs are a problem. This spicy potion will keep your furry friend from becoming your garden's worst enemy. Repeat frequently, especially after each rain.

On the Rise 🏠 FOR REASONS OF THEIR OWN, most dogs avoid raised beds. They want something that's even with the ground where they can roll around. But planting your flowers on the up and up will do more than simply deter cavorting canines. It will also improve soil drainage, and plenty of pesky insect pests hate well-drained soil.

Super Shortcuts

✂ Before you blame a roving canine for unsightly patches in your lawn, take a closer look. Lots of things besides dog urine can cause very similar damage. Your problem could be thatch. Or it could be another culprit, like, well—you! Human error, including gasoline, oil, and fertilizer spills, can cause dead brown spots. And mower blades that are dull or set too low—or a mower that's left running in one place—can do a real number on your lawn, too. The solution: Be careful!

Cheddar and Chicken Bites

Chances are that you love the taste of chicken and Cheddar cheese together. Well, so do your favorite canines. Here's a yummy recipe your dog will gobble up. For a quick flavor change, simply replace the chicken bouillon with beef.

DIRECTIONS: In a large bowl, combine the oats, butter or margarine, and water. Let the mixture stand for 10 minutes. Thoroughly stir in the cornmeal, bouillon, sugar, cheese, milk, and egg. Then add the flour, 1 cup at a time, until a stiff dough has formed. Knead the dough on a lightly floured surface, mixing in additional flour as necessary until the dough is smooth and no longer sticky. Roll out the dough to a ½-inch thickness. Cut it with cookie cutters, and set the treats 1 inch apart on greased cookie sheets. Bake the cookies at 325°F for 35 minutes, or until they're golden brown. Cool before serving. Store the treats in a loosely covered container.

> 1 cup of old-fashioned oats (not quick-cooking)
> ⅓ cup of butter or margarine
> 1 cup of boiling water
> ¾ cup of cornmeal
> 2 tsp. of chicken bouillon granules
> 1 tsp. of sugar
> 1 cup of shredded Cheddar cheese
> ½ cup of milk
> 1 egg, beaten
> 3 cups of whole-wheat flour

A Change in the Wind 🏠 IF YOUR DOG has a peculiar body odor, and he hasn't been rolling around in anything stinky, it could be a sign of an infected wound, decaying teeth, or something more serious. Check your pet thoroughly for any trouble spots.

Q *My dog is getting a bit older, and lately I've noticed that he's having trouble understanding simple commands. This has me worried. Is there any way I can help him get back to his old self?*

A As some dogs age, they can lose their hearing, just like we can. You can help your old buddy adjust by adding simple gestures to your voice commands. For instance, let him know it's time to eat by pointing toward your mouth. Do it consistently, and he'll begin to associate that motion with mealtime.

Cool Treats for Hot Pups

It's hot. It's humid. And everybody—including your canine companion—is as uncomfortable as all get out. What hits the spot on a day like this? A homemade frozen confection, that's what—like one made from this ultra-simple recipe.

Beef bones
Lamb bones
Water

DIRECTIONS: Pick up some beef and/or lamb bones at your local butcher shop (they'll be cheap or maybe even free). Throw the bones into a pot of water, and let everything simmer slowly on the stove for about an hour. Cut any remaining meat off the bones and throw it back into the pot. Let the broth cool completely, then strain it into several ice cube trays, and freeze. When they're frozen solid, transfer the icy treats to ziplock plastic freezer bags or plastic freezer containers. Then, on those hot, sultry summer days when your pup is ready for a treat, toss him one of these while you enjoy a scoop or two of your favorite ice cream.

Play Dress-Up 🏠 IF A DOG'S WOUND isn't bleeding too badly, and you've cleaned it well, you can dress it with the same over-the-counter antibiotic ointment that you'd use on yourself. Apply it with a cotton swab, then keep the wound covered with gauze so it stays clean. Once a day, remove the gauze to wash the wound with warm water, and then reapply the antibiotic cream. Rewrap the cut, making sure the bandage is not too tight.

That's Brilliant!

Just like us, animals can suffer from heat exhaustion and heat-stroke when the temperature shoots up. To protect your pal during hot weather, don't ever leave your pet in the car on a hot day. Many folks just don't realize how quickly a vehicle can become fatally hot—even if you leave a window open.

Delectable Doggie Delights

At first glance, these ingredients may look like the makings of delicious snacks for your human family and friends. Actually, though, they're top-notch canine treats made from first-class "people-grade" food products. And they're just as good for Fido as they are good tasting. (Just don't tell him that.)

2½ cups of whole-wheat flour
½ cup of dry milk
1 tsp. of brown sugar
½ tsp. of salt
6 tbsp. of shortening
1 egg, beaten
½ cup of ice water

DIRECTIONS: Mix the flour, dry milk, brown sugar, and salt in a medium bowl. Cut in the shortening until the mixture resembles cornmeal. Mix in the egg, then add only enough ice water so that the mixture forms a ball. Using your fingers, spread the dough to a ½-inch thickness on a lightly greased cookie sheet. Cut the dough with a dog-biscuit-shaped cookie cutter, and remove the scraps. Pat out the scraps and proceed as before. Bake at 350°F for 25 minutes. When they're completely cool, store the treats in an airtight container.

Perfect Portions ▪ NOT REALLY SURE how much to feed your pal? It all depends on how much he weighs. You can follow these simple guidelines for deciding what size portions to serve your pooch every day: 5-pound dog = ½ cup; 10-pound dog = 1 cup; 20-pound dog = 1½ cups; 40-pound dog = 3 cups; 60-pound dog = 4 cups; 80-pound dog = 4½ cups.

Super Shortcuts

By reducing the amount of hair that your dog will inevitably shed on your couch, you'll also cut down on the amount of time you spend getting all that hair off the furniture. Sound too good to be true? Not at all. The trick is to just brush him! Regular brushing with a good-quality brush is the easiest way to get the loosening hair off your dog before it falls out. So give him a good going-over every day, especially if your pet is long-haired. Trust me—you'll end up spending less time cleaning up the hairy mess, and your dog will appreciate the extra attention.

Dig No More Doggie Spray

It's funny how some dogs have no interest in digging and others tear straight for a flower bed the minute they get outside. If your pooch falls into the latter category, this fabulous fixer has your name written all over it.

DIRECTIONS: Put all of the ingredients into a large jar with a tight-fitting lid, shake well, and let the jar sit for about eight hours. Strain out the solids, then pour the solution into a handheld sprayer bottle. Spray the mixture thoroughly on Fido's favorite digging spots. After one whiff, he'll keep his distance. **Note:** You'll have to reapply the formula every few days and after each rain to keep the scent fresh.

> 1 small yellow onion, finely chopped
> 1 garlic clove, crushed
> 1 tsp. of hot sauce
> ½ tsp. of cayenne pepper
> ½ tsp. of peppermint oil
> 1 qt. of warm water

Test Time 🏠 A SUBSTANCE THAT'S HARMLESS to most critters can cause expensive trouble if your pet happens to be allergic to it. If a switch to natural food doesn't cure your pet's woes, take him to a holistic veterinarian for an allergy test. (It's easier and faster than the kind conventional vets use. It'll cost you a few bucks, but it'll be worth every penny.) To find a holistic veterinarian, visit www.ahvma.org, or call 410-569-0795.

GRANDMA'S OLD-TIME TIPS

One time, when I was little, I really wanted to share my chocolate bar with the family dog. Well, Grandma Putt caught me just in time, and really let me have it. Turns out, too much chocolate might put extra pounds on you, but even one bite can be deadly for your dog. That's because it contains a chemical called theobromine, which can cause severe, life-threatening diarrhea. Baking chocolate is especially dangerous because it's packed with almost nine times more theobromine than milk chocolate has. In fact, as little as 3 ounces of baking chocolate can kill a 25-pound dog. So keep those cakes, cookies, and baking supplies well out of paws' reach!

Dog Be Gone Spray

Dogs may be man's best friends, but they certainly can't be considered a lawn's best pals. When dogs gotta go, they seek relief outdoors— and when they do their business on your turf, you've got a problem. Fortunately, this simple solution can help keep careless canines at bay.

> 3 to 4 hot peppers
> 2 to 3 garlic cloves
> 2 to 3 drops of dishwashing liquid
> 2 gal. of water

DIRECTIONS: Puree the peppers and garlic in a blender, then mix them with the dishwashing liquid and water. Dribble the elixir around the edges of your lawn, driveway, and sidewalks. Reapply frequently, especially after each rain.

Potato Dog Repellent 🏠 GRANDMA TOLD ME that when she was a girl, folks swore by rotten potatoes as a first-class dog repellent. They scattered them in and around new planting beds, outside garbage cans, and anyplace else that they didn't want Rover to roam.

Down, Boy 🏠 KEEP FIDO OFF THE COUCH by laying strips of aluminum foil on the seat cushions. When he jumps up, the crackly sound will startle him, and he'll jump right back down. (At least most dogs will.)

Pet-Hair Magnet 🏠 CAT AND DOG HAIR seems to stick to some fabrics like glue, and even a vacuum cleaner may not do a thorough job of cleaning it up. To get rid of the strays, wipe down your sofa with a damp washcloth. The rough texture of the terry cloth will pull the hairs out of the fabric and roll them up into a snarl that you can simply lift off.

That's Brilliant!

If you have a puppy in the house, you know how he'll chew just about anything once he starts teething. To safeguard your furniture, woodwork, clothing, and everything else within reach, give young Rover a steady supply of big, cold carrots to chew on. The cool temperature will soothe his sore gums, and he'll love the satisfying crunch.

Dog Days Room Deodorant

It's no secret that dogs who spend time romping in the great outdoors (even if it's only your own backyard) can pick up some, shall we say, interesting odors. The next time Rover strolls into the house and leaves the aroma of eau de who-knows-what on your furniture, whip up this marvelous mixer and go to town.

> 2 cups of dried lavender
> 2 cups of dried pennyroyal
> 1¼ cups of dried rosemary
> 2 lb. of cornstarch
> 1 large box of fresh baking soda
> 15 drops of citronella oil
> 2 drops of lemon oil
> 2 drops of pennyroyal oil
> 2 drops of rosemary oil

DIRECTIONS: Put the dried herbs in a blender or food processor and whirl them into a powder. Pour the powder into a bucket, add the cornstarch, baking soda, and essential oils, and mix well. Let the mixture stand in a cool, dark place for two to three days. Mix again and then sprinkle the fabulous fixer anywhere your rooms need a breath of fresh air.

Fit 'n' Trim 🏠 BEING OVERWEIGHT is just as bad for your pet's health as it is for your own. If your dog is packin' a few more pounds than he should be, you could switch him to one of those expensive diet dog foods. But my guaranteed two-step weight-loss program is cheaper, more fun—and a whole lot healthier. First, cut back on the daily dose of dog food, and make up the difference with cooked vegetables. Second, take your pooch for a 2-mile walk every day—it'll whip both of you into better shape.

PINCHING PENNIES

Besides being good for your pet's overall health, fixing your dog or cat can save you money on food bills. That's because neutered males or spayed females need 25 percent fewer calories than their unfixed counterparts. Of course, you'll have to make sure you cut back on the grub you dish out—otherwise, you'll be spending more cash than you need to, and you'll wind up with a chubby Chihuahua or tubby tabby!

Doggie Damage Repair Tonic

> 1 can of beer
> 1 can of regular cola (not diet)
> 1 cup of ammonia

To fix those ugly brown or yellow spots in your lawn caused by dog droppings, remove the dead or dying grass in the affected area. Overspray the turf with 1 cup of baby shampoo per 20 gallons of water, and apply gypsum at the recommended rate. Wait a week, then mix up a batch of this repair tonic to green the grass up again . . . fast!

DIRECTIONS: Mix these ingredients in a bucket, then pour into a 20 gallon hose-end sprayer, and overspray your turf every other week. To prevent future problems, take a stroll over your lawn with a pooper-scooper every day or so, and dispose of the droppings before they can damage your grass.

Whoops! Too Late! 🏠 WHEN THE BURNED SPOTS from dog droppings have been in place for a while, you have only one choice: Dig out the damaged turf, and flush the soil with plenty of water to dilute the salts and nitrogen. Then reseed or resod the renovated areas.

Soak It Up 🏠 TO MAKE IT EASIER to neutralize pet odors in carpeting, remove as much urine as possible while it's still fresh. Start by setting a thick pad of paper towels on the area, then put a section of newspaper on top—and stand on it. You'll be amazed at how much urine the paper absorbs. Keep replacing the paper towels and newspaper until you can't soak up another drop, then rinse the area thoroughly with water.

That's Brilliant!

A couple of nutritious food supplements seem to alter the chemistry of dog urine, making it less damaging to turfgrass. What's more, they help repel fleas. What are these miracle workers? Brewer's yeast and garlic. You can add yeast and garlic to Rover's daily treat menu, or simply sprinkle a little brewer's yeast on his food at dinnertime. **Note:** Check with your veterinarian for the right yeast or garlic dosage.

Doggone Delicious Dog Treats

Here's a pooch-pleasing recipe I stumbled upon in my travels. My dog loves these protein-packed cookie bars, and yours will, too—guaranteed! Just to be on the safe side, though, you may want to show this recipe to your vet to make sure your dog doesn't have allergies or sensitivities to any of the ingredients.

1 lb. of canned dog food
1½ cups of unsweetened wheat germ
1¼ cups of cornmeal*
3 large eggs, beaten
1 tsp. of garlic powder

DIRECTIONS: Preheat the oven to 350°F. Combine all of the ingredients in a large bowl, and mix them together thoroughly. Spread the mixture in a greased baking pan, pop it into the oven, and bake for 20 minutes. Let it cool completely, and cut it into squares. Store the baked treats in a glass jar or other container with a tight-fitting lid.

*Don't use cornbread mix, which contains ingredients other than cornmeal.

Lead a Sheltered Existence

ANIMAL SHELTERS SOMETIMES OFFER pet care like spaying, neutering, and vaccinations at a greatly reduced cost. Call your local shelter to see if it offers those services. And don't overlook pet stores. They often hold clinics and sponsor reduced-fee vaccinations or tests. But, even if you can't find reduced costs, it's your responsibility to make sure these procedures get done for your furry friend's sake.

PINCHING PENNIES

If you live near a university that has a veterinary school, you may be able to get care for all kinds of pets at a fraction of the cost you'd pay to a private practitioner. And that can mean health care, vaccinations, flea dips, and even grooming, so be sure to check it out.

Double-Dip Pet Pleasers

If your household includes both canine and feline family members, this recipe is sure to delight them. Both your dogs and cats will go wild for these flavor-packed goodies.

1 cup of bacon drippings

1 tbsp. of creamy peanut butter

1 cup of ground oats

¾ cup of finely ground, salt-free soda crackers

1 tbsp. of soy powder

DIRECTIONS: Melt the bacon drippings and stir in the peanut butter. Add the oats, crackers, and soy powder, and mix well. Then either wait until the mixture is just cool enough to handle and roll it into balls, or press the still-warm mixture into a baking pan, let it cool to room temperature, and cut it into squares. Store the cookies in a covered container.

Gotta Go! 🏠 IF YOUR DOG IS ALONE most of the day, you may want to consider keeping him in a crate. When he's by himself all day, he may feel lonely or frightened, and that's when accidents can happen. If he spends most of his day in a crate that's big enough for him to stand up, turn around, and lie down in, he'll feel more secure, and a dog is less likely to make a mess where he spends most of his time. Just be sure to give him a good long walk right before you leave and the minute you get home.

Super Shortcuts

✂ If your pet is bleeding, stop the flow by applying direct pressure to the wound with a square of gauze, clean fabric, or handkerchief. The bleeding should stop in a minute or two. If it doesn't, tie a bandage loosely (we're not talking tourniquet here) to hold the cloth in place. If the gauze soaks with blood, don't change it; keeping it in place will actually help the wound to close, and pulling it away may reopen the clot that's forming. If you must, add more gauze or cloth on top until you can get to the vet.

Down-Home Dog Biscuits

Even for a dog, nothin' says lovin' like somethin' from the oven. And you can express your affection for your furry pal loud and clear with this taste-bud-ticklin' recipe that's so easy, Fido could almost bake the biscuits himself!

> ⅔ cup of water*
> 6 tbsp. of vegetable oil
> 2 cups of whole-wheat flour
> ½ cup of cornmeal

DIRECTIONS: In a medium bowl, combine the water or broth and the oil. Add the flour and cornmeal, and mix until a dough forms, adding more flour if necessary. Turn the dough out onto a lightly floured surface, and roll it out to a ¼-inch thickness. Using cookie cutters, cut the dough into shapes, rerolling scraps as you go, and tailoring the size to suit your dog. Transfer the cutout biscuits to a nonstick cookie sheet, and bake at 350°F for 35 to 40 minutes, or until golden brown. Let the biscuits cool on a rack, out of Fido's reach. Store them in a tightly sealed container, where they'll keep for a couple of weeks.

*For even more flavor, substitute chicken or beef broth.

A Lick in Time 🏠 MOST DOGS WILL LICK their wounds. Let them. It's nature's way of allowing them to clean up their cuts and scrapes. After your pet seems satisfied, you can take over. Flush the area with a stream of clean, warm water. And make sure there's nothing lodged in there, like a thorn, splinter, or piece of glass, which may require an appointment with your doggie's doc.

That's Brilliant!

If your dog has a cut paw pad, it's hard to come up with a bandage that the pooch won't tear off the minute you get it on him. So try this: After you clean the wound and cover it with gauze, use a clean, old cotton sock to keep the gauze covered. Just slip your pup's paw right in, pull it up, and loosely tape it on his leg with nonstick first-aid tape.

Fabulous Flea and Tick Shampoo

If fleas or ticks have set up housekeeping on your best pal, give him a bath with this powerful potion. Lots of folks I know swear by it as a first-class way to get rid of the blood-sucking devils.

½ cup of grated castile soap
1½ cups of warm water
2½ tbsp. of glycerin
2 tbsp. of denatured alcohol (use with caution)
3 drops of pine oil

DIRECTIONS: Put the soap and water in a double boiler, and heat until the soap has melted. Then add the glycerin and mix well. Remove the pan from the heat, and while it's cooling, add the alcohol and pine oil. Stir well again. Let the mixture cool completely before using it to bathe your dog as usual.

Hit 'Em Where They Live ⌂ WHEN YOU'RE FIGHTING FLEAS, don't forget to clean all around the area where your pet spends most of his time, mainly pet beds, favorite rugs, and upholstery. Steam-clean the areas if possible, and wash what you can in really hot, soapy water to kill eggs and larvae. Then be sure to vacuum the areas every day to keep fleas from getting comfortable enough to stick around.

Make Fleas Flee ⌂ IF YOU ARE DISPLEASED by fleas, try some strong-smelling herbs to help ward them off. Rub your dog with sprigs of fresh tansy, fennel, or pennyroyal. Use peppermint, lavender, or rosemary sprigs on your cat.

Super Shortcuts

✂ Here's a solution to flea problems that doesn't require a change in your pet's diet or bathing habits. It's a flea comb, and it's available at most pet stores. If your pet has a problem with fleas, you'll initially need to comb his or her coat several times a day. After about a month, a weekly go-through ought to do the trick to keep pesky fleas at bay.

Far Away Feline Formula

It's no secret that cats are territorial creatures with minds of their own. They don't see any problem with digging or rolling in freshly turned dirt—even if it's a spot where you've just spent hours setting out a new groundcover planting! So when your cats—or somebody else's—are up to no good in your garden, safeguard your plants with this potent potion.

> 5 tbsp. of flour
> 4 tbsp. of dry mustard
> 3 tbsp. of cayenne pepper
> 2 tbsp. of chili powder
> 2 qts. of warm water

DIRECTIONS: Mix these ingredients in a bucket or watering can, and sprinkle or spray the solution anyplace you want to set out the "Cats Unwelcome" mat. Any wandering intruders will take one whiff, and then they won't want anything to do with your place anymore. Just make sure to reapply a fresh batch of the formula after heavy rain or watering.

Super-Simple Sprays  WHEN WANDERING TOMCATS leave your garden smelling anything but flowery, relief is as close as your kitchen cupboard. Just fill a handheld sprayer bottle with either undiluted white vinegar or a half-and-half solution of pine cleaner and water, and regularly spritz the places where the boys are leaving their marks. The scent of either potion will confuse and repel the rascals. Just one word of caution: Both of these fabulous fluids can kill plants, so use them only on nonliving things—or on weeds.

PINCHING PENNIES

No cat worth her weight in catnip can resist a nice fluffy bed of freshly tilled soil—like the one where you've planted your annual seeds. Protect those future flowers by crisscrossing the soil with thorny canes. If you grow roses, you can save yourself some money and cut the canes right off the rosebushes. Or, you can get them from bramble fruits.

Fend Off Fleas Dog Treats

These easy-to-make treats pack a potent one-two punch: They deliver a load of dog-pleasing flavor—and they help keep your pal free of fleas. They'll keep for weeks, and they're cheaper than many store-bought varieties. **Note:** A batch of these makes a great gift for any dog owner.

> 3 tbsp. of vegetable oil
> 1 tbsp. of garlic powder
> 2 cups of flour
> ½ cup of brewer's yeast
> ½ cup of wheat germ
> 1 tsp. of salt
> 1 cup of chicken broth

DIRECTIONS: Stir the oil and garlic powder together in a large mixing bowl. In another bowl, mix the flour, brewer's yeast, wheat germ, and salt. Slowly add the oil-and-garlic mixture to the dry ingredients, and stir in the broth a little at a time, mixing thoroughly until you get a doughy consistency. Roll out the dough onto a floured surface to about a ¼-inch thickness. Cut the dough into squares with a knife, or use cookie cutters to make special shapes. Put the treats onto a large greased cookie sheet, and bake at 350°F for 20 to 25 minutes, or until the edges are brown. Let the cookies cool for two hours, then store them in plastic bags or a glass storage jar with a tight-fitting lid in a place where Fido can't get to them.

Frozen Fun 🏠 DURING THE DOG DAYS OF SUMMER, give your dear old pup some welcome relief with a homemade icy remedy. Just fill a large plastic container about halfway with water. Then sink a toy or two—along with a few edible treats—into the water, and put the container in the freezer. Once the treat-filled cube is frozen, plop it into your pooch's food bowl. He'll have a whale of a time trying to get at those goodies—and keep his cool at the same time.

That's Brilliant!

When your dog gets thirsty, you can shoot some water in his mouth. A water pistol is small enough to tuck into your pocket or fanny pack and, let's face it, it's fun to use! Plan ahead—about an hour before you go for your walk, load your "weapon" and put it into the fridge to keep it cool.

Flea-Free Pet Bed Formula

> ½ cup of dried pennyroyal
> 2 tbsp. of dried thyme
> 2 tbsp. of dried wormwood
> 1 tbsp. of dried rosemary

Even after you've gotten the fleas off your dog or cat, it's all but guaranteed that the tiny terrors will still be lurking—and breeding—in his bed. To roust them out, wash the bedcover, or the whole bed if the cover is not removable, then keep them away with this powerful but perfectly safe repellent.

DIRECTIONS: Put the herbs in a food processor and whirl them into a powder. Then open up the cover of your pet's bed, sprinkle the powder onto the pillow, close the cover, and shake the bed to distribute the herbal mixture as evenly as possible. If the cover has a zipper, this maneuver will be a snap. Otherwise, you'll need to carefully open a seam, insert the powder, and sew the opening back up before shaking the bed.

Pest-Fighting Shampoo 🏠 OH, THOSE NASTY FLEAS. What can you do? Well, you don't have to rely solely on harsh chemicals to avoid them. Here's a flea repellent that Grandma Putt liked to recommend to her dog-owning friends. Mix a drop or two of eucalyptus oil in your pet's shampoo. Be careful not to get it into his eyes. Also, be aware of any irritated skin; the eucalyptus might make it worse. The odor, which smells good to us, keeps those pesky fleas away.

Scene of the Crime 🏠 IT'S THE SCENT of previous accidents that tempts pets to relieve themselves near that spot again and again. Our noses aren't nearly as sensitive as theirs are, so to make sure the trail is completely gone, never use ammonia to clean up accidents—it may remove the stain, but its odor encourages pets to reuse the area.

PINCHING PENNIES

There's nothing better than a freshly cleaned pet. And if you're itchin' to get your buddy clean but you're fresh out of shampoo, reach for a super substitute: cornstarch! Just rub a handful into your pet's fur and brush it out. You won't believe how fluffy and spotless his coat will be—with absolutely no shampoo. Or water!

Flee, Fleas Carpet Formula

Whether you've just moved into a new-to-you house and found the carpets infested with fleas, or your own pet has brought the little buggers home, you don't have to put up with them. This marvelous mixer will clear 'em out in no time flat.

> 1 lb. of diatomaceous earth*
> 8 oz. of table salt
> 2 oz. of peppermint powder**

DIRECTIONS: Put all of the ingredients in a bucket and mix thoroughly. Then pour the powder into a container with a perforated lid (like a large saltshaker or a talcum powder container). Shake the powder onto your carpets, paying special attention to the fleas' prime breeding areas—namely the edges, where the fiber meets the wall. Wait for an hour or so, then vacuum. **Note:** Although this formula is nontoxic, the diatomaceous earth can irritate your nasal passages, so try not to breathe in the powder as you apply it, and keep kids and pets off the carpet until it's been well vacuumed.

*Available at garden centers.
**Available at herb shops and online.

Marinated Mice 🏠 MY CAT GOES CRAZY OVER little toy mice stuffed with catnip. She just can't get enough of them—that is, until the catnip aroma wears off, which seems to happen in about a week. Then she wants nothing more to do with the boring things. So I bought a dozen catnippy critters at the discount pet-supply store, and I dole 'em out two or three at a time. The rest I keep "marinating" in a tightly closed glass jar filled with dried catnip. As soon as Kitty shows signs of boredom with her current crew, I pop the "stale" mice into the jar, take out a couple of fresh, odiferous replacements, and watch her jump for joy.

Super Shortcuts

✂ One of the best ways to groom a long-haired critter is with a shedding blade. Pet-supply stores sell 'em in two sizes: small ones for cats and big ones for dogs. The dog versions are exactly the same as the ones used on horses—and you can get 'em at tack shops for about half the price the pet stores charge. Look in the Yellow Pages under "Riding Apparel and Equipment."

Get Things Moving Dog Cookies

Just like you and me, dogs get constipated every now and then. When that happens to your pal, give him a helping hand by making a batch of these delicious, fiber-filled cookies. You'll get things moving again in no time at all!

DIRECTIONS: In a very large mixing bowl, combine all the ingredients. Pour the mixture, a small batch at a time, into a food processor and blend, adding water until the mixture forms a ball. Flatten the dough, roll it out, and cut it into shapes using a knife or cookie cutters. Put the treats on a lightly greased cookie sheet and bake at 350°F for 30 minutes. Cool to room temperature, and store the cookies in an airtight container.

> 1 cup of whole-wheat flour
> 1 bouillon cube, dissolved in ⅓ cup of warm water
> ¼ cup of crushed bran flakes
> ⅛ cup of cornmeal
> ⅛ cup of white flour
> 2 tbsp. of vegetable shortening
> 2 tbsp. of wheat germ
> 1 tbsp. of molasses
> 1 tsp. of sage
> 1 tsp. of salt

Glow in the Dark Pets 🏠 HOW CAN YOU BE SURE that your dog or cat is safe outdoors after the sun goes down? Go to your basement or garage, grab the small reflectors off that old bike, and attach one to your pet's collar. Then if your cat or dog should wander out onto the street at night, passing motorists will be able to spot—and avoid—your beloved pet. And you'll have peace of mind knowing that your four-legged friend is safe no matter what the hour.

PINCHING PENNIES

If your furry friend doesn't like going out in the rain, give him a raincoat! But don't buy one of those fancy pricey ones at the pet store. Instead, make it yourself. Take a trip to your local thrift shop or Salvation Army outlet and head for the children's section. A child's small plastic rain slicker should be just about the right fit for a medium-sized dog. Of course, you may have to do a few small alterations, but it's a heck of a lot better than spending a fortune on designer doggie rain gear!

Home Cookin' Fido Style

When you want to be sure Rover is getting the very best chow, nothing beats Mom's (or Dad's) own home cooking like this recipe.

DIRECTIONS: Brown the meat in the coconut oil. Stir in the other ingredients, let the mixture cool to room temperature, scoop your dog's normal portion into his food dish, and watch him dig in. Store any leftovers in a covered container in the refrigerator.

1 lb. of ground beef

2 tsp. of coconut oil

6 slices of white bread, crumbled

4 cups of cooked rice

2 hard-boiled eggs, chopped

1 canine vitamin tablet, crushed

1 tbsp. of bonemeal

Time to Play ⬆ LONG, REPETITIVE BARKS may mean that your pup's bored. He's trying to tell you that he needs some excitement in his life, so get out there and play with him! Consider enrolling him in obedience or agility training. At the very least, take him to a local park and let him meet and greet other dogs—anything to keep him active and engaged.

Don't Sweat It ⬆ MOVING TO A NEW HOME can be very stressful for pets. So if you're moving and own a dog, find a friend who can keep your pup for a day or two while you get settled. If you must have the dog with you on move-in day, keep his confusion and panic in check by putting him in a securely closed room while you're moving in. Quarantine your cat in this way, too, so she won't be underfoot or scamper out the door. Make the transition even smoother by leaving your kitty or pooch an unlaundered sweatshirt of yours to snuggle up to.

That's Brilliant!

It's true that many tonics you use for grooming and taking care of your dog should be specifically designed for dogs. But if you brush your dog's teeth, you certainly don't need to purchase a special doggie toothbrush. A soft-bristled brush (child size for smaller breeds) works just fine and will probably cost less. Oh, yes, one thing: Just make sure you don't mix up Fido's toothbrush with the rest of the family's!

Infant Kitten Formula

It's a sad fact of life that some kittens are abandoned or orphaned before they're old enough to eat solid food. If you've got a tiny motherless feline at your house (or even a whole litter of them), this formula makes a good substitute for mama's milk.
Note: Check with your veterinarian before you proceed with a full-scale feeding routine.

- 1⅔ cups of unflavored pediatric electrolyte oral solution
- 1½ cups of goat's milk
- 1 cup of plain live-culture yogurt
- ⅓ cup of baby-food lamb
- 2 egg yolks
- 2 tbsp. of white corn syrup

DIRECTIONS: Put all of the ingredients into a blender and mix well. Pour the formula into a pet-nursing bottle and heat it to lukewarm. Test the temperature on the inside of your wrist before feeding a kitten. Store any remaining formula in the freezer until it's mealtime.

Don't Play Hide-and-Seek 🏠 CLOSE ALL OF YOUR closet doors and barricade any other nook or cranny that your new pet could (and, take my word for it, would) get into. Heating vents, washers, dryers, unpatched holes in walls, crawl spaces, and attics are all nifty pet hiding places. And take a look at the undersides of your upholstered furniture; believe it or not, some cats and other small pets find holes in the fabric that covers the bottom and will crawl up into the furniture!

Q *I have a new kitten in my apartment, and I like to leave my windows open. But when I came home the other day I found my kitten with her head poking out the window. What should I do?*

FAQ?

A If you live in an apartment building or even if you have a second story on your house, make sure that your windows stay closed—and if you do open them, check that the screens are secure. Better yet, make it a habit to open windows from the top only, and don't push them down too far.

Just for Love Dog Delights

If there's one thing a dog loves more than getting a cookie for doing something good, it's getting a cookie just because you love him. Your pooch will sit up and beg for these tasty treats—guaranteed!

DIRECTIONS: Combine the all-purpose flour, whole-wheat flour, cracked wheat, cornmeal, dry milk, and brewer's yeast in a large mixing bowl. Stir in the broth, and mix well, using your hands, until the dough becomes very stiff. Gradually mix in the water to get a bread-dough consistency. Roll the dough out on a floured surface to a ½-inch thickness. Cut the dough into shapes, put them on a greased cookie sheet, and brush the tops lightly with milk. Bake at 350°F for 45 minutes, then turn the oven off completely but do not remove the biscuits. Let them sit for at least 10 hours. Store the treats in an airtight container.

- 3 cups of all-purpose flour
- 3 cups of whole-wheat flour
- 2 cups of cracked wheat
- 1 cup of cornmeal
- ½ cup of dry milk
- 1 tbsp. of brewer's yeast
- 2 cups of beef broth
- 1 cup of water
- ⅛ cup of milk

Wall-to-Crate Carpeting

SOMETIMES YOU'VE JUST GOTTA put your dog or cat in a traveling crate, and let's face it, those things aren't exactly cozy. You can always put some old towels in there to make it more comfy, but I've got an even better idea. Stop by the carpet store, and find yourself a scrap or two of carpet that will fit in the bottom of the crate. And since the piece you require will be very small, the merchant just might give it to you for free!

Super Shortcuts

At one time or another, most dogs and cats will put a paw or nose somewhere it doesn't belong. And when that happens, and your pet ends up with a cut or scrape, be sure you tend to it right away. If left untreated, even a mere flesh wound could get infected. So save yourself an expensive trip or two to the vet, and take care of your pet's minor wounds immediately after you notice them. (See pages 230 and 231 for some first-aid tips.)

Lip-Smackin' Luscious Dog Treats

The next time you want to give your best pal an extra-special surprise, bake up a batch of these chickeny cookies. If your dog is anything like mine, he won't be able to get enough of them!

DIRECTIONS: Put the chicken in a medium saucepan, add enough water to cover the chicken, and simmer for 35 minutes, adding more water as necessary. Remove the pan from the heat, and put the chicken in the refrigerator to cool for 20 to 30 minutes. Then slice it into small pieces, and put them in a blender or food processor. Add the remaining ingredients and blend, but do not liquefy. Set tablespoon-size or larger globs on a cookie sheet and bake at 250°F for 40 to 50 minutes. Remove the sheet from the oven and set it in a sunny spot (out of your dog's reach) to dry the pieces out for a few hours. Let the biscuits cool for another 12 hours before storing them in an airtight container.

> 1 lb. of cooked chicken meat (breast, thighs, or combo)
> Water
> 2 cups of old-fashioned oats (not quick-cooking)
> 1 cup of whole-wheat flour
> ¼ cup of dry milk
> 2 large eggs
> 2 tbsp. of soy sauce
> 1¼ tbsp. of garlic powder
> 1 tbsp. of parsley

No, Thanks 🏠 WHEN YOUR POOCH BELLIES UP to the salad bar, make sure he passes right by the raw onions. Some dogs can handle small quantities, but in many cases, a mouthful or more can trigger fever, diarrhea, and vomiting.

PINCHING PENNIES

Almost every pooch I know is crazy about stuffed, plush animal toys. They sure don't come cheap though! You can get around the high pet-store prices by heading for the toy department of your local thrift store, but look carefully before you buy—you want toys that are baby-proof. Otherwise, they may have sewn-on buttons for eyes and noses and other decorations that a pup can easily pull loose. If your dog ends up swallowing pieces of the toy, it could mean an expensive trip to the emergency vet.

Lucky Pup
Leash Conditioner

3½ cups of water
¾ cup of grated Ivory® soap
½ cup of beeswax*
¼ cup of neat's-foot oil**

Do you love your pet so much that you decided to get him a special leather leash just for fancy occasions? Well, that pretty leash can get just as grungy as the one you use every day. Here's an old-fashioned formula for cleaning leather that doesn't need a high polish.

DIRECTIONS: Heat the water to boiling, then reduce the heat to a simmer. Slowly add the soap, stirring gently. Remove from the heat. Mix the beeswax and neat's-foot oil in the top of a double boiler until melted. Slowly add the wax-and-oil combo to the soapy water, stirring until the mixture thickens. Pour the formula into heat-proof containers, and let it cool. To use the soap, rub it onto the leather with a damp sponge, and buff dry with a soft, clean cloth.

*Available at craft-supply stores.
**Neat's-foot oil darkens some types of leather, so test it first in an inconspicuous spot. It's available at most hardware stores and online.

Buckle Up 🏠 IF YOU HAVE an old leather belt that's scratched up or torn, try cutting it and fashioning it into a sturdy dog collar for your best buddy. Just punch holes to fit, and your faithful canine is all set to wear your hand-me-downs!

GRANDMA'S OLD-TIME TIPS

When you find a leather leash or collar for peanuts at a tag sale, don't pass it by just because it doesn't look brand-spankin' new. Just follow Grandma Putt's routine for giving leather a second lease on life: Wipe off any mildew with a solution of equal parts rubbing alcohol and water. Remove any white water spots by covering them with a coat of petroleum jelly. Wait a day, then wipe off the jelly with a soft cloth. Polish up brass buckles with Worcestershire sauce, ketchup, or toothpaste.

Meow-velous Move 'Em Out Mixer

Cats can be just as finicky about what they smell as they are about what they eat. So use this strong-smelling, cat-repelling tonic to convince them that it's not such a good idea to start diggin' in your flower and herb beds, or your newly planted vegetable garden.

DIRECTIONS: Mix the ingredients in a bucket, then dribble it as a scent fence around your no-cats-wanted territory.

*To make tobacco tea, place half a handful of chewing tobacco in an old nylon stocking and soak it in a gallon of hot water until the mixture is dark brown. Pour the liquid into a glass container with a tight-fitting lid for storage.

Turn 'Em On ⌂ INSTALL A BIG BED of catnip as far away from your flower garden as your lot size allows. Most likely, the kitties will have so much fun rolling around in it that they won't give a second thought to your annual and perennial posies.

A More Permanent Repellent ⌂ NO MATTER HOW POTENT a smelly repellent is, you'll have to renew it now and then, as wind, rain, and your garden hose dilute its firepower. Fortunately, there's a longer-lasting solution to your feline frustrations. First, plant seeds of a low-growing groundcover—like creeping thyme or sweet alyssum—around the perimeter of your flower bed. Then lay 2- to 3-foot-wide strips of chicken wire on top of the newly seeded soil. When the plants mature, they'll cover the flat fencing, so you'll hardly notice it. But when Fluffy's paws touch that sharp wire, she'll be outta there!

Super Shortcuts

✂ You can use the same solution that keeps cats out of beds (see "A More Permanent Repellent" above) to keep meowing marauders away from bird feeders and nesting boxes. Just make sure those bird-friendly structures are well out of leaping range of tree limbs, porch overhangs, or other handy platforms. Otherwise, agile feline assassins won't touch down on your chicken-wire deterrent!

Odor-Go Urine Remover

If you've got a dog, chances are that occasionally you'll be surprised by a mess on your carpet. Here's what I use to clean and deodorize areas damaged by pet urine.

> 2 tbsp. of white vinegar
> 1 tsp. of dishwashing liquid
> 1 qt. of water

DIRECTIONS: Mix the ingredients in a bucket, and use a scrub brush to work the solution onto the problem spot. Say, "Out, out, darn spot!" three times and it will be—out in a flash!

Natural Smell Destroyer 🏠 REMOVE THE SCENT of pet urine from carpeting and upholstery the same way Grandma Putt did before those special enzyme cleaners came on the market. If the spot is still wet, blot up as much moisture as you can with paper towels or old rags. Then dampen the spot with water, and pour a generous layer of borax on top. Don't rub it in. Wait until it has dried completely, then vacuum the area.

Hardwood Helper 🏠 WHEN YOUR CAT leaves a puddle on your hardwood floor, try this remedy to kill the odor. Start by blotting up the urine with paper towels. Then mix ⅓ cup of white vinegar with ⅔ cup of warm water and a squirt of dishwashing liquid and scrub the area well. Dry with paper towels, then sponge the spot with a half-and-half mixture of Listerine® mouthwash and hot water, and dry completely.

Q *My cat decided that she couldn't wait to get to the litter box, and instead left me a smelly puddle on my favorite area rug. Is there anything I can do about it?*

A I have a friend whose normally fastidious cat developed a bladder condition (which has since been remedied). Before the problem was diagnosed, though, Kitty let loose several times on a small—and valuable—Oriental rug. My friend was about to toss the rug out, but instead decided to leave it outside for a while just to see what would happen. Lo and behold, after a couple of months in the fresh air and rain, all traces of urine smell had flown the coop. The moral of the story? When all else fails, turn to Mother Nature, and let her do her stuff!

Pleasin' Cheesy Dog Treats

These simple-to-make cookies have more going for them than just pooch-pleasing flavor. They'll also give your best pal healthy doses of protein and fiber.

DIRECTIONS: Cream the cheese and butter or margarine together using an electric mixer. Add the egg and beat the mixture well. Stir in the rest of the ingredients, and mix thoroughly. Cover the dough and chill it in the refrigerator for one hour. Roll it out onto a floured work surface to a ¼-inch thickness. Cut the treats into shapes, and set them on an ungreased cookie sheet. Bake at 375°F for 15 minutes.

*Use any mild-flavored cheese, such as Colby, Monterey Jack, or mild Cheddar.

8 oz. of shredded cheese*
½ cup of butter or margarine
1 large egg
1½ cups of wheat germ
1½ cups of whole-wheat flour
2 tbsp. of milk
Pinch of salt

My Dog Has . . . Gas 🏠 PEEE-YOOO, ROVER! When pets suffer from flatulence, believe me, anyone who's within smelling distance suffers, too! So how do you alleviate this problem? A good brisk walk outdoors should help move the gas out of your pet's intestines. Just make sure you keep yourself upwind.

Tame Tummy Troubles 🏠 DOGS SUFFERING FROM vomiting or diarrhea may not necessarily be ill—instead, they could very well be the victims of their own curiosity! Dogs love to sniff out and sample scraps of food, even if it's stuff that's already been tossed into the trash. So keep your garbage well sealed and out of reach so Fido doesn't get the chance to dine on spoiled food.

That's Brilliant!

A sure cure for doggie gas is activated charcoal, which you can buy at any pharmacy. Check with your veterinarian before giving your pet some of the charcoal (¹/₈ to ¹/₄ teaspoon for small dogs and ¹/₂ teaspoon for larger dogs). Then make sure your pet drinks lots of fresh water. Don't continue this treatment for more than two days.

Pup-Sicles

Keeping your furry friends cool in the summer is just as important as keeping yourself and the rest of your family cool. After all, man's best friend is a part of the family. For hot-weather treats—or to distract a teething puppy from your favorite sneakers—gather some ice cube trays, and make a few batches of these goodies.

> 1 qt. of beef or chicken broth
> ½ cup of finely chopped vegetables
> ½ cup of plain yogurt

DIRECTIONS: Combine all of the ingredients, and pour the mixture into ice cube trays. Then tuck them into the freezer until treat time rolls around. Whenever your pup looks like he could use a cooldown, grab one from the freezer, and watch him wag with happiness as he devours the icy treat.

Doggone! WHEN IT COMES TO DAMAGING your domicile, a cat's claws can't hold a candle to a puppy's sharp teeth. So give the poor baby something to chew on! Just rinse an old, clean washcloth in cold water (make sure there's no lingering soap), and wring it out. Then roll it up tight, and twist it into a sort of spiral. Put it in the freezer till frozen solid, then take it out, and give it to young Rover. But make sure you whisk the cloth away as soon as it's thawed out; otherwise, the pup will recognize it for what it really is and figure that it's okay to chew on all washcloths!

Super Shortcuts

A puppy doesn't have full bladder control till he's almost five or six months old. That's why, no matter how well he understands that his bathroom is outside and how hard he tries to contain himself, accidents still happen. So be prepared with a puppy-training kit. Stock a shoe box or other container with a roll of paper towels, a spray bottle filled with water, a spray bottle filled with hydrogen peroxide, and a pot-scrubbing sponge. Keep this kit handy, and you can take care of any pet mess.

Purr-fect Prevention Potion

If you're looking for a pet that will give you unending love and affection, you'll probably want to steer clear of cats. Cats can be pretty particular about their preferences. But that can be used against them, too. Whip up this potion to send 'em a strong message: Keep your paws out of the garden!

> 1 garlic clove, crushed
> 1 tbsp. of cayenne pepper
> 1 tsp. of dishwashing liquid
> 1 qt. of warm water

DIRECTIONS: Combine the ingredients in an old blender, and puree the heck out of 'em. Sprinkle the resulting liquid around your flower and herb beds, or your newly planted veggie garden. That'll keep the kitties away.

Turn 'Em Off 🏠 YOU CAN PLANT YOUR pussycat protection as an alternative to using cat-repelling sprays in the garden. Just include rue in your planting scheme. It's an easy-to-grow perennial herb with beautiful blue-green leaves, small greenish yellow flowers, and decorative brown seed heads that look great tucked into seasonal wreaths and swags. Best of all, it makes felines flee—fast!

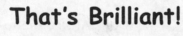

That's Brilliant!

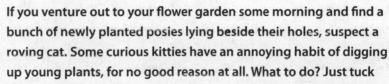

If you venture out to your flower garden some morning and find a bunch of newly planted posies lying beside their holes, suspect a roving cat. Some curious kitties have an annoying habit of digging up young plants, for no good reason at all. What to do? Just tuck the bloomin' babies back into their holes, and give them a good dose of compost to get them growing on the right foot again. Then, to keep Fluffy from trying that trick again, turn up the heat by sprinkling the ground with a spicy water mixture made with chili powder, cayenne pepper, or whatever hot stuff you can find in your spice cabinet. She'll turn tail and run when she gets a whiff of the extra-spicy elixir—and so will just about any other critter that has a nose to sniff with!

Scat Cat Solution

When cats are cuttin' capers in your flower beds or gunning for your fine-feathered friends, put up a "Keep Out!" sign in the form of this zesty potion.

DIRECTIONS: Mix all of the ingredients together, and sprinkle the solution around the perimeter of your yard, or anyplace that Puffy isn't welcome. She'll quickly realize that she should get her kicks elsewhere!

4 tbsp. of dry mustard
3 tbsp. of cayenne pepper
2 tbsp. of chili powder
2 tbsp. of cloves
1 tbsp. of hot sauce
2 qts. of warm water

Chow Time 🏠 MANY CATS ARE perfectly content to use their own litter boxes instead of your flower beds—but they may make the mistake of believing that a container garden is really Joe's Fabulous Kitty-Cat Diner. What do you do in that case? Give the keen kitties something even better to munch on! Most cats prefer wheatgrass or oat grass to garden-variety flowers, paws down. Just get some seed from a garden center, and sow it in a flat or pot that blends with your outdoor decor. When the grass comes up, set the container where Fluffy and Dusty can get at it easily. Then they'll give your posies the cold shoulder.

A Case of Mistaken Identity 🏠 IF YOUR CAT DECIDES to use a laundry basket filled with dirty clothes as a litter box, YUCK! Use an enzyme-containing detergent to dissolve the proteins and the stains, and add 1 pound of baking soda during the wash cycle to neutralize the smell. And while you're at it, freshen up your laundry basket, too, by washing it down with a paste made of baking soda and water.

Super Shortcuts

✂ All across the country these days, more and more folks are doing their gardening in containers. And it seems that cats are mistaking the big pots for their personal potties. Prevent this from happening to your potted posies by saturating some cotton balls in lemon-oil furniture polish, and tucking them into the containers. Kitty won't like the smell one bit! One ball will work for a small container; for larger ones, use two or three.

White Line Disease Soak

1 cup of copper sulfate*
½ cup of white vinegar
8 cups of water

If you're a horse owner, you know that white line disease, a fungus that destroys the inner hoof wall, is a serious problem and can be fatal if it's not treated quickly and properly. Many of my horse-owning friends swear by this formula to head off trouble in the early stages.

DIRECTIONS: Mix the ingredients together in a bucket. Soak each of the horse's hooves in this solution for 15 minutes. Repeat at least five times a week until the fungus clears. **Note:** Have your horse examined by an equine veterinarian before using this or any other homemade treatment.

*Available online.

The Horse in Your Life

BABY THE HORSE in your life by giving him a blanket. No, I don't mean you should put it in his stall so he can cuddle up with it at night! Just spread it across his back before you put on his saddle pad. The blanket will feel soft against his skin, and it'll be a darn sight easier for you to regularly wash than that bulky pad.

Q *I can never find a good way to keep my horse's hooves, mane, and tail shining and sparkling. Is there anything I can do (other than spend a small fortune) to keep these parts looking their best?*

A To shine a horse's hooves, rub in a little hair conditioner, and buff each hoof with a clean cloth. And you can use the same concoction that you use to detangle your own hair on your horse's hair. Just rub a little through the strands of the mane and tail, and comb it out as usual. Your steed will look stunning, and it won't cost more than a few cents.

Winter Wonder Bath

What do you do when Fido or Fluffy needs a bath, but it's way too cold to be strolling around with wet fur? Just reach for this fabulous fixer. Your pet will not only get clean as a whistle without water, but will also stay dry and warm.

DIRECTIONS: Combine all of the ingredients in a container. Rub the mixture into your pet's coat, moving against the direction of the fur. Then brush well to remove the dry shampoo (or use a vacuum cleaner if your pal will tolerate the noise and suction).

> 1½ cups of starch
>
> ¾ cup of unscented talcum powder
>
> ¼ cup of baking soda
>
> 2 tbsp. of borax
>
> 1 tbsp. of cornmeal
>
> A few drops of scented essential oil (optional)

Cat Toys on the Cheap

CATS LOVE TO PLAY and, fortunately, they're easily amused. The best part is that you don't need to spend a small fortune to entertain your fun-loving feline. You can use Ping-Pong™ balls and plastic practice golf balls. They're sold in any sporting-goods store, are cheap, and will keep Kitty amused for hours. In fact, anything a cat can bat around with her paws, like a plastic bottle cap, scrunched-up aluminum foil, or a balled-up sock, will suit her just fine.

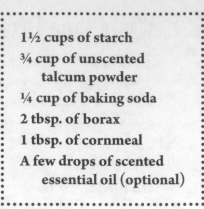

GRANDMA'S OLD-TIME TIPS

Grandma Putt never wasted money on fancy catnip toys for the family cat. She knew that the cat didn't really care what the toy looked like. It's what was on the inside that counted. So Grandma did the cat a favor and put some catnip in the toe of an old, clean sock. She just knotted the sock to keep the catnip in, and let Kitty go wild with her new toy.

CHAPTER 10

BIRD
Buffet

Who doesn't love watching birds flock to their yard? I know I do. I also know all the tips and tricks for luring the widest variety of feathered friends. In this chapter, I'll let you in on my secrets including the best cold-weather foods, how to choose the right feeders and houses, and a roundup of recipes that will tempt blue-birds, orioles, cardinals, finches, tanagers, and many other birds to belly up to your backyard buffet. So get ready to welcome a wave of warblers!

Best Bet for Bluebirds

Like most fruit-eating birds, bluebirds prefer to forage for their feasts—as you know if you've ever watched them raid your cherry tree. In winter, though, when the "real stuff" is hard to come by, you can often lure the blue beauties to your feeder by offering up some of their favorite treats in dried form.

4 parts suet or fat scraps, chopped

1 part peanut butter

2 parts cornmeal, coarsely ground

1 part oatmeal

1 part dried cherries, coarsely chopped

1 part dried currants

DIRECTIONS: Mix the suet or fat scraps and peanut butter together with the cornmeal and oatmeal. Add the cherries and currants, and mix to distribute the fruit throughout. Crumble the mixture, sparingly at first, in a feeder with perching space.

Latch On to a Log 🏠 CHECK YOUR BIRD-SUPPLY STORE for a "suet log," a versatile contraption that serves several birds at one time and blends right in with your natural landscape. They're usually about 1 foot long and 3 inches in diameter, and they're made by drilling holes about 1½ inches across and an inch or so deep into a solid section of a tree branch. You can squish suet or other high-fat foods into the holes, and the birds will cling to the log to eat. If you have a ready supply of logs, you can, of course, make your own version of this wild bird feeding station and use the money you saved to buy more suet!

GRANDMA'S OLD-TIME TIPS

Modern supermarkets usually get their meat pretrimmed of fat, so the scrap supply is much slimmer than it was in Grandma's day. But if you have a neighborhood butcher shop, or a market that takes pride in its custom-cut meat selection, you're in luck! Lots of these places still trim their own meat, leaving an ample supply of leftover fat that's ideal for your birds. If you're a good customer of a small shop, you might even get fat scraps for free.

Blackbird Buffet

Like starlings, blackbirds can flock to a feeder in high numbers and eat the place clean in the blink of an eye. So keep your neighborhood redwings, cowbirds, or grackles content with this cheapie high-fat mix.

4 parts chicken scratch*
2 parts cracked corn
1 part corn oil
1 part suet or lard, melted

DIRECTIONS: Put the chicken scratch and cracked corn in a bowl, and stir to combine. Pour in the corn oil and melted suet or lard, and stir until all dry ingredients are moistened. To serve, crumble about 1 cup at a time directly on the ground or in a low tray feeder. Freeze leftovers in ziplock plastic freezer bags.

*Chicken scratch is a type of chicken feed consisting of whole grain and corn. You can buy it at a tractor-supply store or online.

Yard-Wide Feeding Station 🏠 FROM A BIRD'S VIEWPOINT, your whole yard is a feeding station. Birds often range far and wide to find their favorite foods. By stocking your beds with plants that supply lots of tasty treats in one concentrated area, you'll turn your yard into a magnet for any bird looking for a bite to eat. That's why you'll want to be sure you have plenty of seeds, fruits, and berries—right on the bush!

Your Best Table, Please! 🏠 WHERE YOU PUT YOUR FEEDERS is just as important as what you fill them with. A feeder in a bird-enticing location will quickly fill up with customers, while even the most delectable treats will have a hard time luring birds to a poorly placed feeder. For blackbirds, place the feeder no more than about 5 feet above the ground. Of course, you'll want to put your feeder where you'll have a great view of the action.

That's Brilliant!

Use nuts to add punch to mixes of seeds or fats. As little as ¼ cup of chopped nuts can make the difference between a so-so suet feeder and one that has a line of customers waiting in the wings. Sprinkling ¼ cup of chopped nuts over your seed trays will bring a boom in breakfasting birds, too.

Blue Jay Bounty

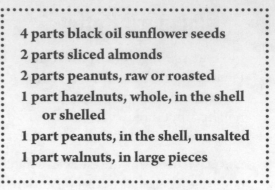

4 parts black oil sunflower seeds

2 parts sliced almonds

2 parts peanuts, raw or roasted

1 part hazelnuts, whole, in the shell or shelled

1 part peanuts, in the shell, unsalted

1 part walnuts, in large pieces

This nutty treat will win the hearts of blue jays, Steller's jays, scrub jays, gray jays, and pinyon jays. Whichever jays you have in your neighborhood, you're bound to see them within minutes of offering this mix. The sliced almonds will slow down these grabby gluttons; they're harder for a jay to pick up with its big beak. Plus, the almonds look funny, so jays may not figure out right away that they're truly a treat worth tasting.

DIRECTIONS: Combine all of the ingredients, and stir until blended. Unless you've won the lottery lately, you'll want to dole this out by offering one or two handfuls at a time. Serve in an open-tray-type feeder to accommodate big-bodied jays.

Jumpin' Jehoshaphat—It's Jays! 🏠 THEY'RE BIG, THEY'RE BEAUTIFUL, they're blue (mostly)—and boy, are they a bunch of bullies! When jays fly in to the backyard, you can bet just about every bird in their path will head for the hills. Jays aren't a danger to other birds at the feeder, but their size and shrieks sure are scary. To prevent those daily panic attacks, I keep a feeder just for jays, stocked with the sunflower seeds and nuts they love best. It helps keep them away from the millet and mixed-seed feeders, where the smaller, shyer birds can eat in peace.

Super Shortcuts

✂ Looking for a way to buy your seed all at once, without having to make an unnecessary number of trips? Find a feed mill. Feed mills and farm stores, often run by farmers' co-ops, may have ultra-low prices on sunflower seed, especially if you live near an area of the United States or Canada where sunflower seed is commercially grown.

Calling All Catbirds

Call in the catbirds with this fruit-rich specialty! They'll really appreciate it in late fall and winter, when natural fruits are very hard to come by.

> 1 cup of suet or fat scraps, chopped
> ½ cup of raisins
> ¼ cup of blackberries, frozen
> ¼ cup of currants
> ¼ cup of fresh apple, chopped
> ¼ cup of strawberries, frozen

DIRECTIONS: Give the bags of frozen fruit a few whacks on a cutting board so that the solid lumps break apart. Combine all of the ingredients in a bowl. Use a large, sturdy spoon to mix lightly so that the fruit and suet or fat scraps are well distributed. Avoid crushing the fruit, and work quickly so that the frozen fruit doesn't thaw. Mold into a block, and serve in a wire suet cage with a perch added.

An Apple a Day

A IS FOR APPLE, and apples are a great beginning for your fruit feeder. Lots of your feeder regulars enjoy an occasional apple, so don't be surprised to see chickadees, titmice, woodpeckers, cardinals, or even juncos taking a nibble. Apples are also aces with the mockingbird and its relatives, the catbird and brown thrasher. Robins and bluebirds won't shy away from apples either.

That's Brilliant!

Raisins are cheap and easy to use, so stock up! But don't just empty that box into your feeder. I've found that birds often overlook raisins when I serve them plain, for reasons I haven't figured out yet. So save these good dried grapes to use in bird recipes, where they'll be eagerly eaten and won't go to waste. I add raisins to recipes to tickle the taste buds of the bluebird, Carolina wren, catbird, brown thrasher, and mockingbird. When I crumble the treats at ground level, I can watch as towhees, sparrows, and robins wolf them down.

Crush on Cardinals

Who wouldn't fall in love watching these wonderful red birds? Cardinals are cold-weather delights at feeders across the eastern two-thirds of the continent. Liven up the gray days of winter with a fluttering flash of red by offering this fruit-flavored treat.

4 parts black oil sunflower seeds
2 parts safflower seeds
2 parts cracked corn
2 parts dried cherries, chopped
2 parts grapes, cut in half

DIRECTIONS: Combine the seeds and corn, stirring until well distributed. Gently stir in the cherries and grapes. Serve in an unroofed tray feeder so cardinals can spot the goodies.

Cherries and Berries Jubilee

SMALL FRUITS ARE BIG temptations for fruit-eating birds. But unless you have your own cherry or berry farm, these luscious fruits are too high-priced to serve up every day. So save them for special treats in the dead of winter or the first days of spring. Of course, fresh cherries and berries cost even more in the off-season (if you can even find them). Not a problem! Just buy them dried or frozen, or process your own when the fresh supply is at its peak. Dry them a pound at a time on a cookie sheet in the oven. An hour or two at your oven's lowest setting or 180°F should do the trick.

Q *Every time I clean out the refrigerator, I find some fruit that's past its prime. Can I feed it to the birds, or should I add it to the compost pile?*

A Unless the fruit is so far gone that it's nothing but mold-covered mush, it's still prime fodder as far as birds are concerned. In the wild, fruit isn't nearly as perfect as the pieces we buy at the grocery store. But there's still plenty of good eatin' in it! Soft, bruised, wormy, frozen—hey, it's all just part of the deal for fruit-eating birds. As long as they can find a few good bites, they'll take you up on that offering of slightly funky fruit.

Downy's Dream

Diminutive downy woodpeckers are among the most reliable feeder guests in any neck of the woods. Treat them well by offering high-fat snack balls made of their favored foods.

DIRECTIONS: Mix all of the ingredients thoroughly, using your hands. Make peanutty balls for an appropriate feeder, or form into a block to fit a wire suet cage.

> 1 cup of suet or fat scraps, chopped
> 1 cup of chopped peanuts
> 1 tbsp. of peanut butter

Woodpecker Powwow

SEE THAT DEMURE DOWNY woodpecker industriously working at your nut-studded suet? Believe it or not, this common feeder visitor has more than 20 woodpecker cousins! Adding nuts to your menu—especially acorns, peanuts, and walnuts—is a good way to make sure you get your share of woodpecker visitors. Serve the nuts whole, or chop and add them to suet mixtures for extra incentive. Depending on where you live, you might catch a glimpse of the fabulous redheaded woodpecker, the striking acorn woodpecker (guess what its favorite nuts are?), and maybe even the eerie white-headed woodpecker.

Crack 'Em Up

WALNUT, HICKORY, OAK, and other nut trees along highways and byways are favorite bird areas when the crop is ripe because the passing cars do a terrific job of cracking the nuts. The crushed nuts are popular with just about every bird around, including sparrows, juncos, and wild turkeys, as well as woodpeckers, jays, and chickadees. Nuts in shells are often available at low prices, so take a tip from crushing cars and smash some nuts yourself. A few whacks with a hammer, and you can enjoy a circus of birds picking through the shells until every bit of nut meat is gone.

PINCHING PENNIES

Nuts vary widely in price depending on the season and the size of the harvest. When the supply of a certain nut is slim, you can expect to pay a premium price. When it's a bumper crop, prices fall. Since birds eagerly eat whatever kind you bring home, make "cents" out of this situation by buying whatever nuts are cheapest.

Finch Favorite

Finches are famous for warbling a few notes while they're frolicking at the feeder. Listen—I think they're singing your praises! The combination of hulled sunflower chips plus seeds with shells will keep finches happily pecking away at this mix. It's a fine feast for goldfinches, house finches, purple finches, and other small songsters.

> 6 cups of black oil sunflower seeds
> 3 cups of white proso millet
> 2 cups of hulled sunflower chips

DIRECTIONS: Mix all of the ingredients together. Pour into a clean, dry plastic milk jug and cap tightly for storage. Use it whenever you see finches flitting around the yard looking for a tasty treat.

It's Feeding Time! IF YOU DON'T SEE BIRDS at your feeders as often as you'd like, perhaps the problem is that they're dining when you're not around. One way to solve this problem is to limit the times when seed is available. Pick an hour in the morning or late afternoon (when birds feed most actively) when you're most likely to have some free time to enjoy a cup of coffee while you bird-watch. Then fill your feeders at precisely that time every day. You'll be amazed at how quickly birds will learn and appear right on schedule!

Q *The little black seeds that my neighborhood goldfinches like so much are sold as thistle seed, niger seed, and, lately, Nyjer™ seed. Now I'm wondering, what's the difference?*

A Only the name. Niger seed and Nyjer seed are the seeds of *Guizotia abyssinica*. Good ol' *Guizotia* is a tall, bushy, yellow-flowered daisy, most definitely not a thistle. I'd wager that "thistle seed" was somebody's bright idea of how to grab the birdseed market because true thistle seeds were widely known to be a favorite of goldfinches. Seed sellers then switched to "niger," an English variation of the name used for the plant in its homelands. Nowadays, "niger" is beginning to give way to "Nyjer," a name trademarked by the Wild Bird Feeding Industry of North America.

Flicker Flapdoodle

You've probably heard a flicker, even if you've never seen one. Just holler its name at the top of your voice, three or four times in a row, fast, and you've got it! Not only does this big brown wood-pecker say its name, it also flickers when it flies, thanks to a white patch of feathers above the tail. Try this suet-rich mix to bring flickers to your feeder.

1 cup of suet or fat scraps, chopped
½ cup of cornmeal
½ cup of cracked corn
¼ cup of nut meats, any kind, chopped
¼ cup of raisins

DIRECTIONS: Put the suet or fat scraps in a deep bowl, pour in the other ingredients, and mix thoroughly. Form into a cake to fit your suet feeder. Serve in a wire suet cage or other suet feeder, secured against a post or tree, so that it doesn't swing under the weight of landing flickers.

Consider it Candy SEEDS AND SUET ARE STAPLES for backyard birds, but nuts are truly treats. They're nice extras, but birds will get along just fine without them, and they'll still visit your feeder even if the nuts are nearly nil. And that's good news for bird lovers on a budget. You can get big pleasure from a small amount of nuts, so don't hesitate to keep a tight rein on your supply!

Super Shortcuts

Double your fun and the amount of birds that can get at your feeder by stringing two or three wire suet feeders together. It'll give the birds more places to perch, and you'll get a kick out of watching the antics as the diners trade places. Just hook the chains of the second and third suet cages to the bottom of the feeder above them. Shorten the chains of the second and third feeders, if you like, by threading them through the cage once or twice before connecting your suet-feeder cascade.

Heavenly for Hummers

This is the gold standard for hummingbird nectar. Heating the water to the boiling point has no health benefit; it simply allows the sugar to dissolve quickly and completely.

DIRECTIONS: Heat the water to or almost to the boiling point. Remove from the heat, and stir in the sugar until it's completely dissolved. Cool, and fill your freshly cleaned feeder. If you've miscalculated the nectar capacity of your feeder, and you wind up with leftover nectar, just store the extra solution in the refrigerator in a tightly capped bottle or jar. It will keep for about three days.

*See Grandma's Old-Time Tips (below) for a caution about the use of sweeteners.

"Gatorade®" for Guzzlers

MIGRATORY HUMMINGBIRDS follow the flowers, traveling north along with the first flowers of spring, and heading south again when blooms decline in fall. That little habit can get them into trouble. If a late-spring cold snap or an early frost catches them in inhospitable climates, your nectar feeder can be a serious lifesaver. At stressful times, boost the sugar-to-water ratio in the recipe above by adding an extra 2 to 4 tablespoons of sugar per cup of water, just to make sure the birds get enough calories to keep on buzzin'.

GRANDMA'S OLD-TIME TIPS

Hummingbirds are so perceptive that they can tell the difference between different kinds of sugar. But that sense of sweet isn't the main reason you'll want to stick to white granulated sugar in your nectar solutions. Honey is just as sweet as sugar, so why not use that? If you'd ever heard Grandma Putt cautioning against feeding honey to infants, you'd know that the anti-honey warnings aren't just an old wives' tale. Honey may contain spores that cause botulism in infants. And it can also pass along a debilitating fungus to hummingbirds. So stick with white sugar!

Mockingbird Mania

Long-tailed mockingbirds are related to catbirds and thrashers, but their personality is much different from their quiet cousins: These guys are just plain greedy! If a mocker lays claim to your feeding station, divert its attention by serving up this recipe on the far side of the house!

> 1 large apple, any kind
> 1 cup of raisins
> 1 cup of suet or fat scraps, chopped

DIRECTIONS: Core the apple, cut it in half, and set it aside. Mix the raisins and suet or fat scraps with your hands until well combined. Using a spoon or your fingers, stuff the hollowed-out center of each apple half with the mixture, piling it on generously. Set the apple treats in an open tray feeder. Or hammer long nails through the back of a board, and push the apples onto the nails. By fastening wire to the two nail heads, you can make a hanging loop and suspend the board anywhere you like in your yard.

Gimme One Good Raisin ... OR HOW ABOUT a whole cupful? Coarsely chop 1 cup of raisins, then set an unwrapped block of room-temperature suet on top of them. Put a dinner plate on top of the suet and press down hard, so the raisins are pushed into the fat. Repeat on the other side of the block. Now you have a treat that will give the jaunty Carolina wren, manic mockingbird, and quiet catbird a good "raisin" to come calling.

That's Brilliant!

Your offerings of fruit will be most eagerly gobbled when they're nearly nonexistent in nature. But in the height of summer, when birds can just as easily raid your garden or nearby orchards, everyone except those good ol' reliable starlings are likely to snub your stuff and get their own. So keep your fruit feeding to a minimum in summer, but ratchet it up as the natural harvest wanes. By September, most fresh fruit is a distant memory to birds, so your supply will be greeted with open beaks from fall through late spring, until natural fruit is once again in season.

Nuthatch Hustle

Natty little nuthatches are always busy. So hustle 'em over to your feeder by offering this delectable concoction.

DIRECTIONS: Thoroughly combine all of the ingredients, using your hands. Serve in a wire suet cage, or liven up the look by forming the mixture into balls for hanging.

> 1 cup of sunflower chips
>
> ¾ cup of suet or fat scraps, chopped
>
> ½ cup of cornmeal, coarsely ground
>
> ¼ cup of peanut butter

Sneaky Steal

PEANUT BUTTER MAY BE PRICEY, but it'll pay off big-time if you live in a neighborhood of feeder keepers. Tempting PB-based treats are a sneaky way to bring the birds over the fence to your place. The stuff is simply irresistible, and the birds will linger until every crumb is gone. Then they'll be back tomorrow to see what you've set out! Of course, the high-fat spread is also packed with good nutrition, so you don't have to feel too guilty about hogging the show.

Chunky or Creamy?

BIRDS WILL GLADLY EAT both types of peanut butter, so you don't need to worry about which to use. Birds that naturally eat nuts will love the chunky version, but will eagerly dine on creamy, too. Those that stick to soft foods may eat around the nuts or swallow those chunky bits right down with the rest of it. It's easy enough to fine-tune your recipes: Since you serve the nut eaters in hanging feeders and the other songbirds in tray feeders, you can give each of them what they like best. But when time is short, or the weather is rough, any kind of peanut butter is better than none at all.

PINCHING PENNIES

Store-brand or generic peanut butters cost a whole lot less than the national, well-advertised varieties. So stick with the cheapies when you're mixing it up for birds, and you can save as much as half the cost on your backyard buffet!

Oriole Ovals

When these sweet singers arrive in mid-spring, it's time to roll out the welcome mat! This treat will keep 'em chirping, and may attract traveling tanagers, too.

> **2 cups of cornmeal**
> **½ cup of orange juice**
> **1 cup of suet or fat scraps, chopped**

DIRECTIONS: Put the cornmeal in a bowl and pour the orange juice over it, stirring to moisten the cornmeal as you do so. Mix in the suet or fat scraps, and mold into balls. Serve in an oriole-accessible feeder, or poke a perch into the ball.

When a Perch Is Paramount

ORIOLES ARE AGILE, but they aren't the winged wonders that hummers are, and their bodies are much larger, too. You'll want to make sure that the nectar feeder you choose for your orioles includes sturdy, spacious perches, so the birds can easily take up a comfortable position and rest easy while feeding. Who doesn't appreciate a comfy seat at the table?

Oranges for Orioles

OH MY, ORIOLES SURE ARE spectacular birds! Their orange or yellow feathers practically glow. It's a good thing there are enough species to cover most of the country, so we all get a share of the beauty. All orioles adore nectar, and in the fruit department, it's oranges that are most appreciated. Service is simple: Just slice the fruit in half and set the halves, cut side up, in an open feeder.

GRANDMA'S OLD-TIME TIPS

When I started putting out my own feeder every year, Grandma Putt told me that most fruit-eating birds are accustomed to foraging for themselves. They can zero in on their favorite fruit faster than you can beat 'em to it. But, when it comes to recognizing their favorite foods at the feeder, birds act like I do when I encounter my dentist in the grocery store: You look familiar, but I can't think of who you are for the life of me! The biggest hurdle in fruit feeding is to get your potential customers to notice the goodies.

Pigeon-Toed Treat

Pigeons (or rock doves, as serious birders call them) and gentle mourning doves are often found among the regulars at feeders in most towns. House sparrows are everyday visitors in backyards, too. Make them all happy with this inexpensive mix.

DIRECTIONS: Measure the ingredients into a plastic pail or large mixing bowl, and stir with your hands until they're blended. Fill your feeder and store the leftovers in a clean, dry plastic milk jug, capped tightly.

> 5 cups of cracked corn
> 4 cups of black oil sunflower seeds
> 2 cups of millet

Old Salts ⬠ I'M FOREVER PREACHING "Hold the salt," whether it's at the bird feeder or at the dining room table. But a giant block of pure salt will quickly get the attention of certain birds, which apparently have never heard of high blood pressure. Natural salt licks were once great gathering places for now-extinct passenger pigeons; their modern kin, the mourning dove and common city pigeon, are just as fond of the salty stuff, which they often glean along the sides of roads that have been salted in winter weather.

The Heat Is On! ⬠ IF SQUIRRELS ARE RANSACKING your bird feeders, turn on the heat to keep them away. Scoop some birdseed into a plastic food storage bag, add a little cayenne pepper, and give the bag a good shake before pouring the seed into your feeders. The squirrels can't stand the heat, but the birds won't even notice!

That's Brilliant!

Both red and white millet are proso millet, with practically no difference except for their color. White proso millet, which is not white but pale golden tan, is the most common millet you'll find. Red millet is a pretty, shiny chestnut color. Birds eat both with relish. But, if you have a choice, choose white over red because birds seem to spot any dropped seeds better, and very little goes to waste.

Robin Roundabout

> 4 parts bread crusts
> 1 part bacon fat, melted

Gentle robins love to pig out on bacon fat, especially in winter, when worms are hard to come by! Prepare this treat only occasionally, so as not to overload them with salt from the bacon. Robins are most likely to visit your feeder for the first time in winter.

DIRECTIONS: Put the crusts in a glass or metal bowl. Pour in the warm bacon fat. Stir gently to coat the bread. Serve in a low tray feeder, where robins can perch around the edges.

Cracker Crumbs 🏠 IF YOU FIND A BOX of crackers with just a few pieces left in the bottom, crumble them into a corner of the seed tray. Don't go overboard though: Except in times of dire need, such as snowstorms, it's best to serve just a scant handful, or they'll go to waste. Experiment with any crackers you have on hand, aside from saltines and oyster crackers. These bland crackers aren't big favorites of birds, so they'll usually go uneaten.

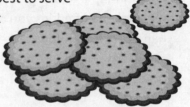

Sticky Snack 🏠 BIRDS LOVE PINECONES stuffed with peanut butter—especially in winter. So gather up several large pinecones and tie a string to the top of each. Spread peanut butter all over the cones, then roll them in sunflower seeds. Hang the treats from your trees, and sit back and watch the birds enjoy their snacks.

PINCHING PENNIES

I know it's a great way to buy food for birds in bulk, but I sometimes get carried away when I'm shopping at the day-old bakery store. After all, who can resist such bargains? Then, when I start to unload my bags of three-for-a-dollar loaves, I realize that, uh-oh, I now have several months' supply of bird bread on hand. That's because the visitors to my busy feeders eat only a few slices of plain bread a day, even in winter. No problem! The extra loaves just go in the freezer.

Sparrow Style

Though not as flashy or as charming as their finchy cousins, native sparrows are just as interesting to observe at the feeder. Serve this spread in a low tray feeder and see how many of these modest characters you can attract. Towhees and red-winged blackbirds enjoy this mix, too.

5 cups of millet

3 cups of cracked corn

2 cups of black oil sunflower seeds

2 cups of flaxseed

DIRECTIONS: Measure the ingredients into a clean, dry bucket or large bowl. Mix well using your hands. Store in a plastic gallon milk jug, tightly capped, until you're ready to put the mix out in your feeder.

Scouting Out Seeds

WHEN YOU'RE READY to buy more birdseed, don't overlook the offerings you can find when you're shopping for yourself. Flaxseed is a bargain at health-food stores, where it's a popular product. Amaranth is a health-food store staple, too, but look for the whole seeds, not the already-ground flour. And don't buy "popped" seeds, which are great people snacks, but not to birds' liking.

GRANDMA'S OLD-TIME TIPS

When Grandma Putt filled our plates for Sunday dinner, she adjusted the serving sizes for each of us. I'd look at the others' heaping plates and beg for more, but she'd shush me and say, "Eat that first, and then we'll see." Usually, her first guess was right, and I couldn't eat another bite. "Eat that first, and then we'll see" is a good guideline when you begin offering pure millet at your feeder. According to those in the know, a pound of millet has about 80,000 seeds. That's a lot of bird bites! And that's why your 10-pound sack of millet will easily outlast 25 pounds of sunflower seed. This little grain is one of the best bargains in Birdseed Land—it serves so many customers with such a small amount.

Strictly for Starlings

Donate your fatty kitchen scraps to a good home—down the gullet of a hungry starling! This low-cost mix will readily feed a crowd of gaping beaks.

> 4 parts cracked corn
> 2 parts cornmeal
> 1 part ham fat trimmings, chopped
> 1 part peanut butter

DIRECTIONS: Put all of the ingredients in a bowl, and combine thoroughly with your hands. The mixture will be crumbly. Scatter it directly on the ground, if you already have starlings visiting. Or set out a sample in a tray feeder, and watch while the birds make the exciting discovery and call in all their relatives to share the feast!

Solid Gold Spread

ONE LOOK AT ALL those bags of yellow cornmeal on my pantry shelf, and you might think I was aiming to win the next Cornbread Cook-off! But those sacks aren't for competitive cooking—they're the backbone of my easiest, one-size-fits-all recipe. I scoop out a few generous globs of generic peanut butter into a bowl, then pour in the cornmeal, ½ cup at a time, mixing after each addition. When the "dough" gets too stiff to mix with a spoon, it's ready for serving. I grease my hands with vegetable oil, so the stuff doesn't stick, and mold some of my mix into patties for my wire suet cages. Then I scoop out the leftovers into a tray feeder or other open feeders for a real feast.

That's Brilliant!

High-fat foods are a must in winter, but they're pretty popular in spring, too. Insects are still in short supply in most places, and spring migrants are streaming back fast and thick. Many of those birds, such as tanagers, bluebirds, and orioles, are soft-food eaters, so they'll be grateful for a generous helping of greasy goodies. Until the spring season is in full swing and lilacs are once again in dooryard bloom, keep birds happy with those high-fat treats!

Tanager Temptations

Consider yourself honored if a glorious tanager shows up at your feeder. This recipe will help get their attention! Orioles and bluebirds may also nab a nibble.

- Half a ripe banana, mashed
- 1 cup of cornmeal
- ½ cup of peanut butter
- ½ cup of suet or fat scraps, chopped

DIRECTIONS: Mix together the mashed banana and cornmeal. Add the peanut butter and suet or fat scraps, and combine thoroughly. Mold into a shape to fit an accessible feeder, or put the mixture in a wire suet cage with a perch poked into it.

Bag It 🏠 WINTER IS PEAK SEASON for suet feeders, so make sure you have plenty of extra blocks on hand. To save time on frosty mornings, stock up on inexpensive wire suet cages before the big rush begins. After you've mixed up some goodies for the birds, stuff the cages with the mix, and slide 'em, cage and all, into a plastic freezer bag. Each suet snack is ready to go anytime! A quart-sized bag is a perfect fit for most wire cage feeders.

It's a Party! 🏠 FRIENDS OF MINE have a fat-fixin' party every fall. One chops the fat, another globs out the peanut butter, a third measures the treats, and the others mix, mold, and package the stuff. In a couple of hours' time messing around in the kitchen, they make enough to stock their feeders for months to come. It's a kid-friendly party, too, because messy, hands-on fun is what these recipes are all about!

GRANDMA'S OLD-TIME TIPS

I never considered trying tropical fruits at my feeder until one balmy May day when Grandma Putt had a few extra bananas going brown. She peeled back a flap of skin, and nestled the 'nanners in the tray beside the apples and fruit treats. Was I surprised when a scarlet tanager fluttered down to take a taste! Tropical fruits, I learned, are just the ticket for our summer birds that winter in the Tropics.

Thrashers' Thrill

And it's a thrill for you, too, when you spot an imposing, long-tailed thrasher at your bird buffet. This mockingbird relative is a solitary type, so you're likely to see only one at a time.

> 4 parts suet or fat scraps, chopped
> 2 parts cornmeal, coarsely ground
> 2 parts raisins
> 1 part peanut butter
> 1 part fresh dark purple grapes, any variety, cut in half

DIRECTIONS: Mix the suet or fat scraps, cornmeal, raisins, and peanut butter in a bowl, combining thoroughly. Mold the mix into large, flattened balls. Now for the tempting touch: Poke the grape halves partway into the suet balls so they're highly visible to grape-eating thrashers. They'll dig right in!

Bird-Tempting Trail Mix 🏠 HEY, CAN I HAVE a date? Thanks! I'll add it to this trail mix I'm cooking up for my tanagers and thrushes. Birds eat dried fruits at the feeder with just as much relish as they do fresh ones, so I regularly patrol the bulk-foods aisle of my supermarket and natural-foods stores to see what's on sale. I've found great buys on chopped dates, apple slices, currants, apricots, peaches, and other goodies, including that classic, dried plums, a.k.a. prunes. It's all fair game for mixing with chopped suet and molding into bird-attracting treats!

Q *I kept hearing about how much most birds love fruit, so I went and put grapes in my feeder. Problem is, the birds don't seem to be eating them. What's going on?*

A The smaller the grapes, the better for birds, because they like to gulp down these juicy fruits whole. If you're lucky enough to live near wineries, or have a farmers' market nearby, you may be able to scout out grapes in smaller sizes than some of those monsters at the supermarket. But if supermarket grapes are your only choice, you'll find birds can eventually figure out how to eat them. Just give them some time to learn how, or wait for the prime customers to show up—the mockingbird, catbird, and brown thrasher.

Titmouse Time

In England, *tit* means a small bird. This one's as soft and gray as a mouse, and a sprightly favorite at the feeder. So try this mix to give those titmice something delicious to snack on when they come around.

> 1 cup of cornmeal
> 1 cup of peanut butter
> 1 cup of sunflower chips
> ¼ cup of cracked corn

DIRECTIONS: Combine all of the ingredients in a bowl, and mix thoroughly. Pack the mix into a wire suet cage, and then sit back to watch the show.

Twice the Titmice 🏠 INSTANT GRATIFICATION is almost guaranteed when you put up a birdhouse for titmice. These little gray birds, with a feathered crest on their heads that makes them look like miniature cardinals in spite of their color difference, are so quick to appreciate new digs that they're likely to start moving in as soon as you move out of the area.

Early Birds 🏠 TITMICE WILL TELL YOU when it's time to get out there with the hammer and nails! They're among the earliest birds to begin springtime courtship, tuning up with love songs by late winter. When you hear a loud, clear, whistled "Peter! Peter!" (easy and fun to imitate!), you'll know it's titmouse time. By Valentine's Day, the titmouse twosome is acting like a pair of newlyweds, and beginning the hunt for a hospitable home, where they may start raising a brood as early as March.

PINCHING PENNIES

Look on the lowest shelf at your supermarket to spot the giant-sized jars of peanut butter. But before you stock up, check the shelf tags and see what the price per ounce is, compared to smaller sizes. I was surprised to find out that sometimes the biggest jar just isn't worth it—its price per ounce is exactly the same as the smaller, easier-to-handle sizes! But when the giant jar is the cheapest option, I take it home. Then I just transfer its contents to smaller jars I've saved, so I don't have to wrestle with a heavy 5-pounder every time I want to make up a mix.

Woodpecker Welcome

We all know when a woodpecker is in the neighborhood. That speedy tap-tap-tap is a sound that even a bird novice can identify right away. Delight your downy woodpecker and his larger cousins with this nutty, corny mix. Chickadees, titmice, and nuthatches will approve, too!

> 1 loaf of bread, any kind
> 2 cups of suet or fat scraps, chopped
> 1 cup of cracked corn
> ½ cup of walnuts, chopped

DIRECTIONS: Tear the bread into pieces in a large bowl. Add the suet or fat scraps, the cracked corn, and the walnuts. Mix and mold firmly with your hands to fill two wire suet cages.

A Deeper Look at Custom Fit

SHOPPING FOR A WOODPECKER HOUSE is like buying clothes for a big family. You may have to visit the children's department, the juniors', and maybe even the big-and-tall section, depending on what kind of woodpecker calls your neck of the woods home. Not only do woodpecker requirements differ for the size of the entrance, but the depth of the box is significant, too. So once you figure out what kind of woodpeckers are in your neighborhood, you'll have to choose your house accordingly.

Q I thought I read somewhere that feeding bread and cake to birds is bad for their health. But I see people tossing bread to the birds every time I go for a walk in my local park. Are they killing the birds with kindness?

A There's no need to cut the carbs because these foods aren't the staple of any bird's diet. They'll get plenty of more nutritious items to balance any of the things they grab a bite of at our feeders. The sugar in breads, cakes, or muffins isn't "bad" for birds—it just adds more carbs to fuel their daily activities. But there is one ingredient to beware of: chocolate. Substances in chocolate can actually cause harm to animals by constricting blood vessels and overstimulating the heart.

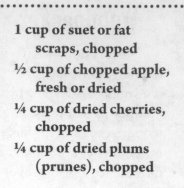

Wrangle a Wren

The big, bright-eyed Carolina wren seems to be changing its habits in recent years, and sticking around for the winter season. Help keep this friendly bird stay in fine fettle with this fatty, fruity recipe.

DIRECTIONS: Combine all of the ingredients, using your hands to mix the fruits with the fat. Stuff a wire suet cage with the mix.

> 1 cup of suet or fat scraps, chopped
>
> ½ cup of chopped apple, fresh or dried
>
> ¼ cup of dried cherries, chopped
>
> ¼ cup of dried plums (prunes), chopped

An Iffy Proposition 🏠 PUT UP A WREN HOUSE in your yard, and a house wren will come. But whether it stays or not is another story! The male house wren builds several trial nests before the female arrives to select her favorite. So, as you watch a bird carefully poke twigs through the hole, keep in mind that it isn't a guarantee that babies will soon be on the way. But even if your birdhouse isn't selected as a permanent residence the first time around, it may be perfect for a later batch of chicks, or even next year's nesting season.

Super Shortcuts

✂ To save myself a trip to the store, I buy apples by the bushel because they're one of my main anti-starling devices. No, I don't throw the fruit at them! I use apple halves to lure those ever-looting starlings away from my other feeders. When times are tough in winter, even starlings get my sympathy. So I set out a half dozen apple halves on the ground, far away from my feeders. The starlings will happily peck every last bit of flesh from the apple, until nothing is left but a thin, empty shell. Farmers' markets and orchards are a great place to save money on apples. Look for the less than perfect ones, sold at a bargain price. Birds won't mind the blemishes one bit.

CHAPTER 11

PEST
Peeves

Pests come in all shapes and sizes, from biting mosquitoes to odiferous skunks, and they can make life miserable. But you don't have to turn to professional exterminators to combat these cunning critters. In fact, most of my pest repellents are made from simple ingredients that you already have on hand—and they won't cost you an arm and a leg. So turn the page and see what's in my arsenal—you'll find just the recipe for banishing pesky pests from your home and yard.

273

Ant Ambrosia

If there's one group of uninvited guests that even indoor picnics can do without, it's ants. Having a line of ants leading into the house, or a pile of ants clinging to something sweet in the garage, is even worse (and it looks really creepy, too). To keep local ant armies in check, try this mixture.

> 4 to 5 tbsp. of cornmeal
> 3 tbsp. of bacon grease
> 3 tbsp. of baking powder
> 3 packages of active dry yeast

DIRECTIONS: Mix the cornmeal and bacon grease into a paste, then add the baking powder and yeast. Dab the gooey mixture on the insides of jar lids, and set them near the anthills. Before long, the lids will be full of ants that met their untimely demise. Keep replacing the lids until they aren't filled anymore, and your kitchen (or other area in the house) will be ant-free again.

The Two-Day Wonder 🏠 TRAPPING THE WORKERS will ease your ant troubles for a while. To get rid of a colony for good, though, you need to eliminate the brains behind the brawn: the queen, who's hunkered down in her bunker doing nothing but churning out eggs by the thousands. And here's an easy way to do it: Just sprinkle instant grits on top of the anthill. The worker ants will carry the grains into the nest, where they and Her Majesty will have a feast. Then the grains will swell up inside their little bodies. Within 48 hours or so, the whole colony will be history.

That's Brilliant!

While most ants are perfectly content in the great outdoors, some prefer cozier quarters—like your home. Well, just because they do good work in the garden doesn't mean you have to offer them hospitality in your kitchen! Try laying sprigs of fresh mint where the little fellows are coming and going. The ants'll go back to where they came from.

Ant Elimination Elixir

1 cup of sugar

3 cups of water

1 tbsp. of boric acid powder

Ants are possibly the most annoying of all pests (well, except for that nosy next-door neighbor). They show up everywhere, and always in large numbers. So say "Adios, ants!" with this powerful tonic. Beware, though: Boric acid is toxic, so don't use this formula if you've got children or pets on the premises.

DIRECTIONS: Add the sugar to the water in a saucepan and bring to a boil. Then stir in the boric acid. Pour the solution into small jar lids, and set the lids in the middle of ant trails or near anthills. Store any unused portion in a secure container, and keep everything out of reach of children and pets. **Note:** When you're preparing or handling this bait, wear rubber gloves and protect your eyes, nose, and mouth with goggles and a mask.

Into the Drink
IF YOU'RE FRESH OUT OF GRITS (see "The Two-Day Wonder" at left), reach for this old tried and true ant-control weapon: boiling water. Just scrape the top off the anthill, and quickly pour a kettle full of the steaming liquid into the nest. And act fast before the workers swarm all over the place! If the water reaches its target, the queen will be an instant goner, and any workers who survive will soon die of old age. Check back in a week or so; if the colony still shows signs of activity, treat 'em to another boiling-hot shower.

GRANDMA'S OLD-TIME TIPS

When Grandma saw ants getting into the house, she devised a way to use some of the plants she grew in the garden to remedy the situation. Once she got the ants out of the house, she kept them from coming back by planting tansy, spearmint, pennyroyal, or southernwood in the vicinity. The little rascals will keep their distance from all of 'em!

Beat the Ant Blues Batter

Alum*
Borax
Flour
Sugar
Water

Ants aren't always a bad thing to have around. After all, they're just doing their jobs, trying to find food for the colony. But when the little rascals are making themselves at home in your home, you have no alternative: It's time to turn to this potent problem solver.

DIRECTIONS: Mix equal parts of the dry ingredients with enough water to make a batter. Pour it into shallow pans, and set them out where the ants congregate—but where children or pets can't get to the traps. Then step back and you'll see the pans fill with the nasty little buggers, leaving your home ant-free. With any luck, it'll stay that way.

*Available in the spice aisle of your supermarket.

Teatime for Ants 🏠 GOT ANTS IN YOUR HOUSE? No problem! Just use dried mint to brew up a batch of strong tea, and spray it on the ants' pathways. The little rascals will turn right around and go back to where they came from!

Say Adios with Orange 🏠 KILL INVADING ANTS instantly by spraying them with an orange-scented household cleaner. Then prevent their relatives from staging a comeback by spraying a line of the same cleaner across any entry point to your house. That'll show 'em who's boss.

That's Brilliant!

When the ants in your yard are driving you crazy, put a piece of tape over the hole in the bottom of a flowerpot, and set the pot upside down on top of the anthill. When the little fellas emerge from the nest, they'll scramble up the sides of the pot. Then all you have to do is pick it up and drop it into a bucket of boiling water.

Bee Balm Buzz-Off Solution

Does the faint *bzzz, bzzz, bzzzing* of mosquitoes hovering on your window screens drive you buggy? Here's bee balm to the rescue! First, wash the screens with a mild soap. Then spray on my bee balm infusion.

> 3 cups of water
> 2 cups of dried bee balm flowers and leaves
> 1-qt. glass jar with a lid

DIRECTIONS: Bring the water to a boil in a saucepan. Crumble the flowers and leaves and place them in the jar. Pour the boiling water over the mixture, bringing the water level to about an inch from the top of the jar. Stir the herbs with a wooden spoon to make sure they're well soaked. Put the lid on the jar and let the potion steep for about 15 minutes. Then strain out the herbs and pour the liquid into a handheld sprayer bottle. Now you're ready to take up the battle against unwanted pests. Just be sure to spray the screens from the inside out so you don't make a mess on the floor.

Ring the Dinner Bell 🏠 BATS, FROGS, TOADS, TURTLES, lizards, ants, praying mantises, spiders, dragonflies, and birds (especially purple martins) all eat mosquitoes. So lay out the welcome mat for the happy snackers! It's easy—create a habitat that includes a water source, plenty of shade, and welcoming structures like bat houses and bird feeders.

FAQ ?

Q *My husband is itching to install an electric bug zapper on our backyard patio. But I'm skeptical—will it really work to reduce the mosquito problem in our outdoor dining area?*

A Yes and no. Those gadgets zap bugs—just about every kind except mosquitoes! That's because most of the zappers use light to attract their "prey." And mosquitoes don't care beans about light. Those suckers are out for blood, so they zero in on the smell of carbon dioxide, which mammals give off. What you want is a machine that uses CO_2 to lure skeeters to their deaths.

Bug-Free Window Wash

½ cup of bay essential oil
¼ cup of white vinegar
3 cups of water

It's a treat to turn off the air-conditioning and leave the windows open on a cool summer's evening. But not when you hear a constant whine of insects trying to get in! Here's an easy way to keep mosquitoes, flies, and other insects from staging a "we-want-in" convention at your windows.

DIRECTIONS: Put the bay oil, vinegar, and water in a handheld sprayer bottle. Tighten the nozzle and shake well. Spray the mixture on your windows inside and out to clean them and make 'em sparkle! And the best part is that the bugs won't want anything to do with your windows after you've used this mixture.

And Stay Out! 🏠 YOU'VE SPENT GOOD MONEY stocking your pantry, and a lot of time putting it in order. The last thing you need now is a parade of multilegged munchers marching into the cupboards and helping themselves to your stash. So how do you keep the varmints out? Simple: Scatter some bay leaves across the shelves before you restock them. Wily weevils, beastly beetles, and other pesky pests will keep their distance.

Knock Their Socks Off! 🏠 HERE'S A GREAT USE for old socks that have lost their partners (but only use socks without holes). Fill 'em with cedar shavings and tie the tops shut. Then tuck 'em in drawers among your clothes and linens to fend off moths and silverfish.

That's Brilliant!

If you've ever painted the exterior of your house in warm weather, you know what a pain it is to find bugs in your paint. If you don't want dive-bombing mosquitoes and other insects to embed themselves in your wet house paint and give your siding and trim a look that you weren't exactly going for, try this simple trick. Just add a couple of drops of citronella oil to every gallon of paint. It won't affect the paint, but it will keep bugs out of your paint job.

Classic Roach Killer

¼ cup of cooking oil
or bacon grease

¼ cup of sugar

8 oz. of boric acid
powder

½ cup of flour

½ cup of chopped
onion

This fabulous fixer is one of the most popular roach killers of all time. And for good reason—it works like a charm! Beware, though: Boric acid is toxic, so don't use this formula if you've got children or pets on the premises.

DIRECTIONS: Combine the oil or bacon grease and sugar in a bucket or old pot. Mix in the boric acid, flour, and onion, adding more oil as necessary to get a dough-like consistency. Shape the soft dough into marble-sized balls, and put them in open plastic sandwich bags (the bags help keep them moist longer). Distribute the open bags around your home, in the corner seams of your windows and doors, near tiny floor or wall cracks, or wherever roaches enter, as well as their favorite hangouts—kitchen cabinets, drawers, under the sink, and other dark, damp areas. When the dough gets as hard as a rock, whip up a fresh batch of bait balls . . . unless, of course, the first batch wiped out the problem completely. **Note:** When you're preparing or handling this bait, wear rubber gloves and protect your eyes, nose, and mouth with goggles and a mask.

Bait Those Bad Bugs! 🏠 SOCK IT TO INVADING roaches with a homemade trap. Simply apply a liberal glob of petroleum jelly all around the inside of a glass jar, then partially fill the jar with some kind of bait, like beer-soaked bread, bacon grease, or pieces of fruit. Wrap the outside of the jar with a paper towel, and set it where you've seen the bugs. They'll crawl up the paper towel to get to the bait, fall into the jar, and then won't be able to get back out!

PINCHING PENNIES

Wondering what to do with all that bacon grease left over from your Sunday brunch? Just drop dollops of the stuff into shallow containers, like jar lids or cat food cans, and set them around your garden. Earwigs will flock to the stuff. Then all you'll need to do is pick up the traps and dump them into a bucket of soapy water.

Cockroach-Conquer Concoction

4 tbsp. of borax
2 tbsp. of flour
1 tbsp. of cocoa powder

When you're thinking of an absolutely disgusting insect, the lowly cockroach usually comes to mind. Most folks consider them to be the ultimate sign of dirtiness. But, contrary to most assumptions, cockroaches can invade even the cleanest of houses. If they've started to move into your domain, don't call an exterminator. Instead, get the pests outta there with this simple formula.

DIRECTIONS: Mix the ingredients together, and put the mixture into jar lids. Set them in your kitchen cupboards, behind the refrigerator, and anyplace else where the roaches are roaming—but be sure that you put 'em in places where children or pets can't get to the chocolaty bait.

Mosquito-Bite Helper ⬠ THE NEXT TIME A MOSQUITO sticks its bloodsucking snout into your skin, nix the itch and swelling by dabbing the spot with a few drops of ammonia. Act fast, though, before you start scratching. If you apply ammonia to broken skin, the sting will feel a whole lot worse than the skeeter's bite!

The Itsy-Bitsy Spider ⬠ COBWEBS—THEY'RE THE MOST DREADED sight in my house. It's no secret that spiders and I have never been on friendly terms, so you'd best believe I get rid of them as soon as I see that first string dangle down from the ceiling. To get cobwebs out of hard-to-reach places, simply put an old sock on the end of a yardstick, and sweep it around the ceiling. Voilà!

Super Shortcuts

When you spend a lot of time in a flower garden, you're bound to find yourself on the wrong end of a bee now and then. And if you're smart, you've got more than a few tricks up your sleeve for easing the pain and swelling. One of my favorite bee-sting remedies comes straight from the laundry room. After scraping out the stinger, just dab a few drops of bluing onto the spot for instant relief. You can find bluing in some grocery stores or online.

Double-Duty Roach and Ant Repellent

> 1 cup of borax
> ¼ cup of crushed bay leaves
> ¼ cup of crushed fresh black pepper

When roaches and ants are dealing out double trouble, you could go to the store and buy two different pesticides. Or you could reach for this versatile mixer. It'll polish off both kinds of pests—and keep them from coming back.

DIRECTIONS: Pour all of the ingredients into a jar with a tight-fitting lid and shake well. Sprinkle the mixture into the corners of cupboards and drawers, and anyplace else the demonic duo are causing trouble.

Talcum Powder Line THERE ARE CERTAIN LINES that ants won't cross, no matter how much they want the tasty treats on the other side. For instance, ants won't cross a line of talcum powder. So use it at the entrance to your pantry, or wherever else you don't want the tiny troublemakers to trek.

Earwig Eliminator EVEN THOUGH THEY LOOK threatening, earwigs rarely cause real trouble. Sometimes, though, the chomping bugs can get out of hand. If that happens at your place, pour equal parts of vegetable oil and soy sauce in empty cat food or tuna fish cans. Set out the traps at night, and toss them out (along with their contents) early in the morning before butterflies or good-guy bugs drop by for a drink.

GRANDMA'S OLD-TIME TIPS

Are anthills making a mess of your lawn? Use Grandma's favorite anti-ant solution to bid 'em adieu. Simply mix ¼ cup of liquid hand soap and 1 gallon of water in a bucket, and pour the solution on the mound. Repeat the procedure about an hour later to make sure the liquid penetrates to the queen's inner chamber and takes out the production factory at its source.

Easy Roach Exterminator

Roaches love the combo of sugar and chocolate just as much as humans do. They'll flock to these taste-tempting traps—and the other ingredients will send them to the big roach motel in the sky.

Note: This formula is toxic, so don't use it in places where children or pets can get to it.

> 2 lbs. of 10X powdered sugar
> ½ lb. of borax
> 1 oz. of sodium fluoride*
> ½ oz. of cocoa powder

DIRECTIONS: Put all of the ingredients in a bucket and mix thoroughly. Then sprinkle the powder in all of the problem areas inside your home. Store any leftovers in a clearly labeled closed container, well out of reach of children and pets.

*Available at drugstores.

Chill Out 🏠 CHILL THE PAIN of a bee sting with ice. Put ice cubes or crushed ice in a cloth or a plastic bag, and place it on the sting for 15 to 20 minutes. You'll soon say *ahhhh,* and the pain will disappear like magic.

Be a Web Master 🏠 TO PREVENT SPIDERS from webbing up your windows, spray rubbing alcohol on the sills. If you don't want to run the risk of the alcohol ruining your windowsills' finish, you can scatter a handful of perfumed soap shavings on the sills instead. Either method will deter spiders from spinning their webs, and save you from cleaning up after them.

Super Shortcuts

Do you have a problem with roaches and other nasties in your kitchen? Don't fret; simply sprinkle whole cloves in any areas where you have pest problems. It's a safe and effective repellent that'll keep roaches and other bad-guy bugs away. Just remember to replace the cloves every few weeks, so they always remain potent.

Farewell, Fruit Flies Formula

Pesky fruit flies hover around your fruit bowls and potted plants and drive you downright batty! So get the upper hand by luring them into a tempting trap that's custom-made to the tiny terrors' taste.

> 2 tbsp. of wine (any variety)
> 2 tsp. of hot water
> 5 drops of dishwashing liquid

DIRECTIONS: Mix the ingredients in a small glass bowl, and set it out where the gnats congregate. Every fruit fly in the room will be drawn to it and take the deadly plunge.

Rhubarb Spray 🏠 THIS MAY SEEM like an odd thing to say, but I was always secretly overjoyed whenever aphids or whiteflies got out of hand in Grandma's garden. How come? Because I knew that she'd probably go at 'em with her rhubarb spray. She'd pluck enough leaves to make about 1 pound and boil them in 4 cups of water for about half an hour. She'd strain out the solids, and pour the liquid into a handheld sprayer bottle. Then she'd mix in 2 teaspoons of dishwashing liquid and spray those little bugs to Kingdom Come. After that, she'd go back to the kitchen and use the stems to make my favorite strawberry-rhubarb pie.

Shoot 'Em Down 🏠 TRYING TO TAKE DOWN flying insects? Don't even think about using toxic bug sprays. Instead, mix a squirt of dishwashing liquid and ¼ cup of white vinegar in a 32-ounce handheld sprayer bottle, filling the balance with water. Then you can just spray 'em dead in their tracks, er, flight path.

That's Brilliant!

If you have a problem with codling moths in your apple tree, make sure you take into account this little bit of information. When the first infected fruit drops, pick up every single piece and destroy it. That way, you'll make a sizable dent in the next generation. Whatever you do, don't let that wormy fruit get anywhere near the compost pile, or you'll soon have BIG trouble!

Flying-Pest Potpourri

When you invite guests over for a summertime get-together, the last thing you want—and the one thing you'll probably get—is a swarm of flying insects joining the festivities. Well, whether you're letting the good times roll indoors or out, this aromatic fixer will say a loud, clear "Bug Off!" to uninvited party guests.

DIRECTIONS: Combine the ingredients in a large container and blend well, using either your hands or a big spoon. Pour the herbal mixture into baskets or bowls, and set them around the scene of the action. You and your guests will love the aroma—and flying pests will hate it!

*Available at craft-supply stores, herb shops, and nurseries that specialize in herbs.

8 yellow tulips, dried

1½ cups of dried lavender flowers

1½ cups of dried rosemary

1 cup of dried southernwood

½ cup of dried pennyroyal

½ cup of dried santolina

½ cup of dried spearmint

¼ cup of cedar wood chips (as fresh as possible)

¼ cup of mugwort

¼ cup of tansy

3 tbsp. of powdered orrisroot*

Give Bugs a Wash 🏠 I'VE HAD GOOD LUCK removing stuck-on dead bugs from my car with a mixture of 3 parts glass cleaner (the blue kind) with 1 part dishwashing liquid. Let it sit for a minute, and then wipe off the bugs with a sponge. If there are some stubborn hangers-on, use a nonabrasive plastic scrubbie to remove the residue.

PINCHING PENNIES

The most effective way to say "Good riddance!" to fruit flies in your trees is with sticky-ball traps. But don't buy them at the garden center. It's easy to make your own, and better yet, let the kids or grandkids do it—it's a great rainy day project. Before the blossoms on your trees turn to fruit, buy some red apples with the stems still attached. Spray them with a commercial adhesive, or coat them with petroleum jelly, and hang them in your trees. Those orbs will soon be covered with flies that thought they'd found the perfect maternity ward. Cut them down and replace them with new ones.

Fruit-Fly Fixer

The fruit you've brought home from the farmers' market is ripe and delicious, but what about the annoying fruit flies that stowed away in your purchases and have now invaded your kitchen? Put 'em to rest with this handy trap.

> ⅔ cup of water
> ⅓ cup of white vinegar
> 1 tsp. of honey

DIRECTIONS: First, stow your fruit in the fridge; don't attract more fruit flies by keeping it on the counter. Then mix all of the ingredients in a rinsed out 20-ounce soda bottle. Use an ice pick to poke holes in the sides of the bottle above the liquid, and leave the bottle where the fruit flies have been buzzin' around. They'll find their way into the bottle in no time, but won't be able to get out. Keep the bottle out until you don't see any more activity.

Drown and Out 🏠 TO GET RID OF irritating indoor fruit flies, fill a small glass halfway with apple cider vinegar and 2 drops of dishwashing liquid. Mix the solution well, then set the glass away from areas where people gather. The annoying flies will be drawn to the glass, fall in, and drown.

Get 'Em While They're Eating 🏠 ONCE FRUIT FALLS from the tree, it doesn't take maggots long to burrow into the soil and start pupating. So act fast! Pick up and destroy fallen fruit as soon as you spot it on the ground, from summer through fall. If you can't patrol your yard daily, at least make an inspection tour a few times a week.

Super Shortcuts

✂ As a fallback measure to any other attempt to get rid of fruit flies that are attacking your grove, use an organic mulch around your fruit trees, or plant a dense ground cover. Either ploy will invite ground beetles and rove beetles to set up housekeeping, which is a good thing because they gobble up fruit-fly maggots as well as the pupae.

Grandma Putt's Foolproof Flypaper

Every summer at the start of fly season, my Grandma Putt would hang up strips of sticky flypaper to trap the pests on the wing. Well, those fabulous fixers work just as well today as they did back then. What's more, these homemade strips are perfectly safe to use around children and pets.

¼ cup of maple-flavored syrup (not the expensive pure stuff!)
1 tbsp. of brown sugar
1 tbsp. of white granulated sugar
1 brown paper bag
Rimmed baking sheet or shallow pan

DIRECTIONS: Mix the syrup and sugars in a small bowl. Cut the bag into strips about 8 inches long by 2 inches wide, and poke a hole about an inch from one end of each strip. Lay the paper pieces in the baking sheet or pan so that they're not touching each other. Pour the syrup and sugar mixture over the strips and let them soak overnight. Then tie a loop of string through each strip (or use S hooks), and hang the traps wherever flies are prone to flit.

Stinging Meat Tenderizer 🏠 DO YOU KEEP meat tenderizer in your kitchen? Well, it's perfect for treating insect bites. Just mix it with a few drops of water and spread it on the stricken site. The tenderizer will break down the protein in the poison, nixing its pain-producing properties.

GRANDMA'S OLD-TIME TIPS

Contrary to their name, clothes moths don't confine their activity to clothing. Any natural-fiber object in your house—including your carpet—can be a moth maternity ward. If you suspect that larvae may be lingering in your floor covering, do what Grandma Putt did to put an end to their buffet. Saturate a bath towel with water, wring it out, and spread it over the rug. Then grab your iron, set it on high, and press the towel until it's dry. The steamy heat will kill the pesky little pests.

Lose 'Em with Lard Roach Remover

There are plenty of traps and tricks for getting rid of roaches, but this old-time formula is one of the best. **Note:** Boric acid is toxic to humans and animals, so don't use this bait if you have children or pets on the premises.

> 8 oz. of boric acid powder
> ½ cup of all-purpose flour
> ¼ cup of lard
> ⅛ cup of sugar
> Cool water

DIRECTIONS: Combine the boric acid, flour, lard, and sugar in a bowl, and mix thoroughly with enough water to form a dough. Then tear off small pieces of the dough and roll them into balls. Put one to three dough balls into small open plastic bags and set them in the problem areas. The roaches will scamper in to eat the "cookies"—and they won't get out again. **Note:** When you're preparing or handling this bait, wear rubber gloves and protect your eyes, nose, and mouth with goggles and a mask.

Clobber Cockroaches 🏠 THESE OBNOXIOUS PESTS are infamous for the mischief they cause in kitchens and other indoor rooms, but they also invade garden sheds and workshops. Fortunately, it's a snap to get rid of them—without resorting to possibly dangerous pesticides. Simply make a half-and-half mixture of sugar and baking powder, and sprinkle it over the infested territory. The bugs will scurry to gobble up the sugar, and the baking powder will kill them. Replace the supply as needed.

Q *My wife swears she saw a roach in our kitchen last night, but I've never spotted one. I'd like to nip this potential problem in the bud, but how can I tell if we even have a problem?*

A Roaches are nocturnal, preferring to live and eat in the dark. They spend daylight hours in moist, dark places, like behind the refrigerator or under the sink. So you can try to spot 'em while they're snoozing. And be on the lookout for their droppings, which resemble coffee grounds or black pepper.

Mighty Mouse Control Mix

Here I come to save the day! Sure, that little mouse with the cape was a lot of fun to watch. But his real-life counterparts aren't quite as heroic. To keep mice and other small varmints from nibbling your young fruit trees and shrubs to nubs, give 'em a taste of this highly effective repellent.

> 2 tbsp. of cayenne pepper
> 2 tbsp. of hot sauce
> 2 tbsp. of Murphy® Oil Soap
> 1 qt. of warm water

DIRECTIONS: Mix the ingredients together in a bucket, then pour the liquid into a handheld sprayer bottle. Drench all of your young tree and shrub trunks with it in late fall. To keep the scent fresh, repeat every few weeks and when the weather is rainy.

Foiled Again! 🏠 BEFORE WINTER SETS IN, wrap the trunks of your shrubs and young trees loosely in aluminum foil to a height of 18 inches to 2 feet. The glittering, rattling surface will send mice looking for food elsewhere!

Anti-Mouse Mints 🏠 WHEN IT COMES TO MUNCHABLES, there's nothing mice like more than car insulation—the stuff on the underside of your hood and between the passenger compartment and the engine wall. And there's nothing mice hate more than mint. So put this secret to work for you by saturating a few cotton balls with peppermint oil and tucking them in your car's insulation. I guarantee that Mickey, Minnie, and their pals will find less aromatic quarters before you can say "Ears to you!"

That's Brilliant!

As much as they dislike hot scents, mint is one cool aroma mice hate. They'll flee if you plant it among your shrubs, but there's an easier way to put mint power to work in your yard. Just mix 2 tablespoons of peppermint extract or peppermint oil in a gallon of warm water. Pour the solution into a handheld sprayer bottle, and thoroughly spritz the crown and lower stems of your shrubs. To protect container plants, simply saturate a cotton ball in the oil or extract, and tuck it into the pot at the base of the plant (for large specimens, use two or three balls).

Move On, Mosquitoes Mixer

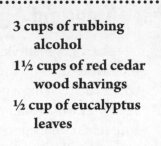

3 cups of rubbing alcohol

1½ cups of red cedar wood shavings

½ cup of eucalyptus leaves

Nothing puts a damper on a summer day—or evening—like throngs of marauding mosquitoes. Well, I've got good news for you: Whether you're headed for a hike in the woods or simply eating dinner on your deck, this marvelous mixer will keep the bloodsuckers at bay. Bonus: It smells good, too!

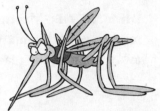

DIRECTIONS: Put all of the ingredients into a large bowl and mix well. Cover the bowl and let it stand for six days. Then strain out the solids and pour the liquid into a handheld sprayer bottle—or several pocket-size bottles if you plan to take the repellent on an outing. Spray it on your skin as needed (at least every couple of hours).

Itch Relief 🔊 WHEN MOSQUITO BITES have you itching like crazy, don't scratch 'em—just reach for a bottle of Listerine®. Moisten a tissue with the mouthwash, hold it on the bite for about 15 seconds, and kiss that ol' itch good-bye.

Don't Give Blood 🔊 ONE OF THE EASIEST WAYS to repel mosquitoes is to serve up a menu that includes a lot of fresh garlic, which seems to keep the biters at bay. And why not? After all, they are vampires. So add a little extra garlic to whatever it is you're cooking (you can't go wrong with the additional flavor), and you'll be able to ward off plenty of the nasty little pests.

Super Shortcuts

You probably can't eliminate every single low spot in your yard and garden, but you can make sure none of them becomes a skeeter nursery. When you find a body of water that you can't empty, pour in some oil (vegetable, mineral, or neem). It'll spread across the surface and smother the future biters. (Even though mosquitoes spend their infancy in the water, they have to breathe air to survive.)

Safe-and-Sound Insect Spray

There are plenty of pest-control sprays on the market that do their jobs effectively. But when you're using a pest-control spray inside your home on your houseplants, you want to make sure that it's harmless to people and pets. Well, look no further—effective insect repellents don't come any safer than this one.

DIRECTIONS: Mix the oil and baking soda in a small bowl. Then combine 2 teaspoons of the mixture with the water in a handheld sprayer bottle, and fire away when you're ready to send those houseplant pests packin'.

Tick Repellent THE SAME MENTHOLATED RUB that's been easing chest colds since Grandma's day is also a terrific tick repellent. Just smooth the stuff onto your skin, and the tiny disease-spreading terrors will give you the cold shoulder.

Ta-Ta Tick! TO REMOVE A TICK from your skin, put a blob of petroleum jelly on it and leave it alone. After about 30 minutes, the tick will loosen its grip, and you'll be able to wipe the little bugger right off or grab it cleanly with a pair of tweezers.

GRANDMA'S OLD-TIME TIPS

As the old saying goes, variety is the spice of life. And no one knew that better than my Grandma Putt. She knew that variety in a flower garden can also help you avoid a whole lot of pest problems. That's because most plant-eating insects have definite food preferences. When they look down on a big patch of their favorite vittles, they'll drop in for a feast. But when they see a lot of different plants—some good, some so-so, and some they wouldn't touch even if they were starving—they generally take their appetites elsewhere. So, when you shop for plants, do yourself a favor and buy as many different kinds as your space allows.

Say Nuts to Gnats Potion

Put a stop to those annoying gnats that sneak through your window screens with this easy-to-make elixir. You can use any of the following perennials: bee balm, common yarrow, southernwood, tansy, or mugwort.

> 3 cups of fresh flowers and or leaves
>
> 3 cups of boiling water

DIRECTIONS: Put the flowers and/or leaves in a large bowl or pot, and pour the boiling water over them. Cover and let sit for 30 minutes. Strain, then spray the remaining liquid on your window screens (from the inside). You'll be amazed at how quickly those gnats scatter, and how long they stay away. Just reapply it anytime the buggers seem to gather in large quantities again.

Go Fish! IF YOU HAVE A POND or water garden, you probably don't want to drain it or cover it up to keep mosquitoes from taking it over, and you shouldn't have to. Just stock it with goldfish, minnows, or guppies. They're all first-class skeeter eaters. If you're still up to your ears in buzzers, though, go for the big guns: Call your Cooperative Extension Service, or a nursery that specializes in water gardens, and ask where you can get some gambusia. These little fish never grow

more than 2½ inches long, but they polish off so many mosquito larvae that, in many parts of the country, fish hatcheries are breeding them just for that purpose.

That's Brilliant!

When you invite friends to a barbecue, you want 'em to *have* dinner, not *be* dinner. To dine in peace, try setting out an oscillating fan or two—mosquitoes don't like moving air. Or toss a handful of sage, rosemary, or citrus peels on the coals of your grill. You'll keep the biters at bay and spice up the chow at the same time!

Scoot, Skeeters Solution

Who wants to be swatting away at mosquitoes when they should be enjoying the great outdoors by relaxing in their favorite patio chair? I know I sure don't. So do what I do and keep those bloodsuckers away on warm summer nights with this tonic.

> 1 cup of lemon-scented ammonia
>
> 1 cup of lemon-scented dishwashing liquid

DIRECTIONS: Pour the ingredients into a 20 gallon hose-end sprayer, filling the balance of the sprayer jar with warm water. In the early morning or late evening, thoroughly soak any places around your yard where mosquitoes tend to gather. Repeat this process three times each week, and it will keep the little buggers away. This tonic works nicely for the rest of the yard, too.

Lemon Mosquito Repellent 🏠 THERE ARE JILLIONS of mosquito repellents on the market these days, but for my money, none of them can beat the ones that Grandma grew in her herb garden: lemon thyme, lemon balm, and lemon basil. To put them to work, all you have to do is crush the leaves to release their volatile oils, and rub them on your skin. You'll love the strong, citrusy scent—but skeeters will avoid you like the plague!

GRANDMA'S OLD-TIME TIPS

Back when Grandma Putt was fighting mosquitoes in her yard, they were garden-variety nuisances—not the vile disease carriers they are today. But although the stakes may be higher now, Grandma's battle plan works as well as ever to keep the population down. Set some old pans around your lawn, fill them with water, and add a few squirts of dishwashing liquid to each one. When the mama skeeters set down to lay their eggs, they won't be able to get up again. Better yet, when the eggs hatch, the larvae will drown!

Sensational Silverfish Repellent

Silverfish can destroy anything that's made of paper. So when you put your precious books, photographs, files, or artwork into storage, tuck one of these mixers into each box. The tangy aroma will make ravenous rotters want to keep their distance.

DIRECTIONS: Mix all of the ingredients in a large bowl, then divide the mixture into muslin bags, cheesecloth pouches, or panty hose toes. Tuck one bundle into every box that's going into storage. To be extra-safe, especially if you live in a damp climate, put pouches behind books that are sitting on shelves, and scatter some around your kitchen, too. The herbs' pungent aroma will keep silverfish far, far away.

All that Glitters May Be . . . 🏠 SILVERFISH. THESE LITTLE CRITTERS love damp areas, and they feed on organic matter like paper. So when you store books and magazines or fabrics like cotton and wool in a damp garage, you might as well start ringing the dinner bell for these pests. Be sure to put anything that could potentially become infested in sealed plastic bags, or wrap it tightly in plastic. While you're at it, include a small amount of mothballs or moth flakes around the packaging for added insurance.

That's Brilliant!

Silverfish do their dirty work at night, and during the day they like to hide out in cracks and crevices to wait for darkness to fall. So use your vacuum cleaner attachment a few times a week to take aim at any possible lairs. Then use the little nasties' nocturnal nature against them, and keep a few lights brightly lit overnight in places where you've seen them show their ugly mugs.

Skeeter Send-Off

I've come across a whole lot of mosquito-repellent recipes in my day, and plenty of them work just fine. One of my favorites is this potent herbal mixer. When you rub it on your skin, I guarantee the skeeters will stay away in droves!

½ cup of dried chamomile
½ cup of dried nettle
½ cup of dried pennyroyal
¼ cup of sweet basil
1 cup of peanut oil
½ cup of sweet orange oil

DIRECTIONS: Crush the herbs. Pour the peanut and orange oils into a double boiler and add the crushed herbs. Heat, stirring occasionally, for about 45 minutes. Cover the pan, remove it from the heat, and let the mixture cool. Strain the mixture through a fine sieve, mashing the herbs to acquire the most fluid (and the most herbal aroma) possible. Store the elixir in a well-sealed container in the refrigerator until you need it. Then when you're headed for skeeter territory, pour some of the oil into a plastic cosmetic bottle, and tuck it into your pocket or purse. Rub the oil onto your exposed skin every couple of hours, and you'll be skeeter-free for the duration.

Coffee Drowning

IF YOU HAVE LOW POINTS in your yard that fill with water after a rain (or when your sprinklers go off in the morning), then you've got a breeding ground for mosquitoes. To take care of them, pour some cold leftover coffee into the puddles. Forget the decaf, though; it's the caffeine that does the trick. When mosquito larvae are exposed to caffeine, they get so confused that they can't tell which end is up, and they drown!

Super Shortcuts

Need a potent outdoor insecticide right now? This one will kill some of the peskiest pests around, including aphids, leafhoppers, cabbage worms, and even mosquitoes. Just mix 1 tablespoon of garlic oil and 3 drops of dishwashing liquid in 1 quart of water in a handheld sprayer bottle, take aim, and spray your troubles away.

Skunk-Away Fabric Formula

> 1 qt. of white vinegar
> 1 cup of baking soda
> 1 tsp. of grease-cutting
>   dishwashing liquid

Skunk "perfume" contains an oil that makes it cling to your pet's fur—and then of course it spreads to your carpets, upholstery, and anything else your pet comes in contact with. So when your home has a close encounter of the skunky kind, sink the stink with this potion.

DIRECTIONS: Pour the ingredients into a handheld sprayer bottle, put on a pair of rubber gloves (so the skunk oil doesn't rub off on your skin), and thoroughly spray the smelly spots. Blot up the liquid with an old dry towel, and continue spraying and blotting the area until the eau de skunk is gone. If you can't get the stink to budge with this formula, you may need to call on a professional to steam-clean your carpeting and upholstery.

Skunked! 🏠 YIKES! YOU FOUND YOURSELF at the wrong end of a skunk—and now you and your clothes stink to high heavens! Here's what to do: Immediately strip off your clothes and dump them in the washing machine with your regular detergent and 1 cup of white vinegar. Then head to the shower. Scrub yourself from head to toe, using plenty of body soap and shampoo. And if you still can't stand yourself, mix up a batch of Skunk-Away Fabric Formula (above) in a plastic bowl, reducing the baking soda to ¼ cup. Bring it into the shower with you, and use a washcloth to completely wipe yourself down with the concoction.

GRANDMA'S OLD-TIME TIPS

Grandma Putt was not afraid of anything, including tangling with a skunk. She relied on a certain trick to tackle a skunk before it had a chance to spray. When a skunk first wanders onto your turf intent on setting up housekeeping, a good spray of water can send him scurrying. Just grab your garden hose, stand at least 20 feet away from the trespasser, and blast away.

Skunk Odor-Out Tonic

Skunks generally spray their foul liquid when they feel threatened. But at night when you're sound asleep, you might not know exactly what threatened a skunk. The only thing you do know is that you have to face the morning with the lingering telltale stink. When a skunk comes a-callin' and leaves some fragrant evidence behind, reach for this easy remedy.

> 1 cup of bleach or white vinegar
> 1 tbsp. of dishwashing liquid
> 2½ gal. of warm water

DIRECTIONS: Wearing gloves, mix all of the ingredients in a bucket and thoroughly saturate exterior walls, stairs, or anything else your local skunk has left his mark on with this potent potion. **Note:** Use this tonic only on nonliving things—not on pets or humans. The bleach and soap could be ingested, requiring immediate medical attention.

Fence 'Em Out 🏠 SKUNKS RARELY CLIMB FENCES, which gives you an easy and attractive way to safeguard your territory. For the surest protection, use ¼-inch wire mesh that's 3 feet wide. Then, to keep the rascals from tunneling under, bend the bottom 6 inches outward at a 90-degree angle, and bury it at a depth of 6 inches. To disguise the 2 feet left above the ground, just tack it to a picket fence or other decorative yard accessory.

PINCHING PENNIES

If you'd rather not erect a fence to keep skunks out of your garden (see "Fence 'Em Out" above), just cover your planting bed's surface with hardware cloth. Make sure the openings are no larger than ¼-inch square, and don't smooth it down flat. The more waves you leave in the surface, the less the critters' sensitive little feet will like it. To improve the appearance, you can spread a thin layer of soil or mulch over the metal. But be sure to keep the layer light so the little buggers can feel those sharp strands.

Sweet and Spicy Moth Chasers

If you hang a pomander ball in every closet in your house, you'll keep moths away like nobody's business.

DIRECTIONS: Combine the cinnamon, orrisroot, and nutmeg in a bowl and set it aside. Take 1 foot of the ribbon and cut it in half. Tie one of the strips around the orange and knot it on top. Do the same thing with the other strip of ribbon, so that the two strips crisscross the orange. (To make sure the ribbon doesn't slip, stick pushpins or thumbtacks in the spots where the two strips cross each other.) Next, stick the cloves into the orange so that they cover the entire orange—but not the ribbon. Then put the clove-studded fruit in the bowl with the spices, and gently roll it around. Let it sit in the bowl for five days, turning it occasionally. Take the orange out of the bowl, hold it over the sink, and, using a hair dryer on a low, cool setting, blow off any excess spices. Attach the remaining ribbon to the pomander ball, and hang it from the closet rod.

*Available at craft-supply stores, herb shops, and nurseries that specialize in herbs.

> 1 oz. of ground cinnamon
> 1 oz. of powdered orrisroot*
> ½ oz. of ground nutmeg
> 2 ft. of ¼-inch to ½-inch-wide grosgrain or other non-silky ribbon
> 1 thin-skinned orange
> 1 box of whole cloves

Clean 'Em Up 🏠 I'LL SAY IT AGAIN: Clean all of your winter clothing before you transfer it to storage. You've heard about moths eating wool, right? Well, technically, it's not the moths that do the damage; and they don't actually eat the wool. Moths lay their eggs on the wool, and when those eggs hatch, the little larvae eat the dirt on the wool. If there's no dirt on the clothing, you won't have to worry as much about moths.

Super Shortcuts

A safer alternative to toxic naphthalene mothballs or flakes is to use cedar wood chips or dried lavender flowers. They'll keep pests away—and they sure as heck smell a whole lot better!

Sweetly Lethal Ant Traps

1¼ cups of active dry yeast
1¼ cups of sugar
½ cup molasses or honey
20 small plastic lids or bottle caps

There are few things more frustrating than seeing a line of ants marching into your home, sweet home. Fortunately, I've got a honey of a fixer that'll stop the little buggers in their tracks.

DIRECTIONS: Put the yeast, sugar, and molasses or honey in a bowl, and mix thoroughly until the consistency is smooth. Spoon a small amount of the mixture into the lids or caps, and set them near ant trails. Bingo: End of ant antics!

Chalk It Up ANTS GET IN just about everywhere when they want to. If they're raiding your garden shed, keep them out by sprinkling a powdered chalk line around exterior doors and window frames. The troublemakers won't cross the line.

Cucumber Deterrent WHEN GRANDMA PUTT peeled fruits or vegetables, she tossed most of the skins onto her compost pile. But for a few of them, she had other plans in mind. For instance, Grandma used her leftover cucumber peels to repel ants. Just lay the peels in their pathways, and they won't come around anymore.

GRANDMA'S OLD-TIME TIPS

Grandma Putt was a live-and-let-live kind of person, even where pests were concerned. But one fall, when ants began making first-class nuisances of themselves in her kitchen, she fought back hard. And you can, too. Just mix 2 tablespoons of sugar and 1 tablespoon of active dry yeast in 1 pint of warm water. Then spread the mixture on pieces of cardboard, and set them in the problem areas. The ant invasion will be finished in a flash.

Ultra-Safe Anti-Flea Mixer

½ cup of baking soda
½ cup of cornstarch
½ tsp. of rosemary, pennyroyal, or citronella oil*

If you have a pet that's contracted an itchy little problem, it's only a matter of time before that problem infests your carpet. Here's an old-time, ultra-safe way to get rid of fleas that hitch a ride indoors with Fido—and keep your carpets clean and fresh-smelling at the same time.

DIRECTIONS: Before you go to bed, mix the ingredients together. Spread the mixture evenly across your carpet, and work it into the fibers using a stiff brush or broom. Let it sit overnight, and vacuum it up in the morning.

*If fleas are not a problem and you simply want to clean your carpet, substitute any fragrant oil of your choice.

Carpet Flea Cleaner 🏠 TO MAKE FLEAS FLEE from your carpet, sprinkle it with a mixture of baking soda and table salt. Let it sit overnight, then vacuum well. Repeat the procedure two more times on dry-weather days, and you'll be flea-free! If you live in a damp climate, sprinkle the mixture in the morning, and vacuum a few hours later.

Dry Flea Remover 🏠 IF FIDO'S FLEAS HAVE MOVED into the house, sprinkle borax onto the carpet, then vacuum it up. (Be sure to get the borax into all of the cracks and crevices in the floor and woodwork, too.)

That's Brilliant!

Is it flea season again? Kill the tiny terrors by pouring salt into all the cracks in your wood floors and baseboards. To make a carpet flea-free, salt it at a rate of 6 pounds of salt per 100 square feet of rug. Wait 24 hours, and scrub with strong soap or carpet cleaner.

Y'all Go!

Even harmless bugs can make monumental nuisances of themselves when they're flitting around your head—or your dinner table. Well, this excellent elixir will build a barrier against pesky flying pests—as well as their more harmful cousins like fleas and mosquitoes.

> 4 oz. of sweet almond oil
> 25 drops of citronella oil
> 10 drops of cedarwood oil
> 10 drops of eucalyptus oil
> 10 drops of pennyroyal oil

DIRECTIONS: Mix the oils together in a glass container with a lid. Douse cotton balls with a few drops of this elixir, and dab it on countertops, windowsills, and other places that are prone to bug invasions. Bugs absolutely hate the scent! When you're heading outdoors for a walk or hike, dab a small amount of this herbal blend onto any exposed skin. Just be careful not to get the oil near your eyes. You'll smell like a million bucks and be able to enjoy the great outdoors without being bugged!

The Nose Knows 🏠 OR DOES IT? Flea beetles find their targets by scent, and most of the beetles are host-specific. That gives you an easy way to keep these pesky pests out of your flower garden: Forget beds, or even long swaths of only one kind of plant. Instead, go for a crazy-quilt mixture—the more, the merrier. Those itty-bitty beetles will be so confused by all of the different smells that they'll just give up and go somewhere else for dinner.

Super Shortcuts

Believe it or not, aphids can't stand bananas. So, if you love these tasty fruits as much as I do, you've got a never-ending anti-aphid arsenal. Just lay the peels on the ground under your plants, and aphids will keep their distance. As an added bonus, these skins will break down and enrich the soil with valuable potassium and phosphorus. Now that's what I call a win-win solution!

CHAPTER 12

HOUSEPLANT
Hoedown

If you're like me, your home is filled with happy, healthy houseplants. On the other hand, maybe some of your indoor greenery is looking a bit, shall we say, tired. Well, don't fret. Follow my tips and tricks in this chapter to perk up your plants, feed 'em right, and keep pesky pests away. You'll be rewarded with lush, healthy indoor greenery that will brighten up the long, dark days of winter and make your home decor delightfully inviting any time of year.

Crystal-Clear Vase Cleaner

Having fresh-cut flowers in a beautiful crystal vase can add a splash of color and class right where it's needed most in your home. But crud builds up quickly inside crystal vases, especially if you've left old flowers in them a little too long. Freshen yours up with this terrific tonic.

DIRECTIONS: Mix the ingredients, and pour them into the vase, making sure the liquid fills the container to the top. Let the mixture sit in the vase overnight, then in the morning, dump the solution and hand-wash the vase with soapy water. Rinse it, let it dry, and fill the sparkling beauty with a fresh bouquet.

Clean That Vase 🏠 WE ALL KNOW HOW GUNK builds up in the bottoms of crystal vases, cruets, and decanters. Instead of trying to scrub it out with a brush, take a tip from the folks who create the most expensive crystal: Fill the vessel about halfway with water, and pop in two denture-cleaning tablets. Let it stand for an hour or two, then rinse well.

Wave Good-Bye to Water Lines 🏠 WHEN HARD WATER leaves lines on your favorite vase, do what Grandma did: Saturate a towel or washcloth (depending on the size of the container) with white vinegar, and stuff it in so it has contact with the sides. Let it sit overnight and by morning, those unsightly marks will wash right off.

GRANDMA'S OLD-TIME TIPS

One day, Grandma Putt accidentally cracked her favorite ceramic vase. But she didn't get upset—in fact, she didn't bat an eye. She just grabbed a block of paraffin from her jelly-making cupboard, melted the wax, and poured the hot liquid into the vase. Then she rotated the vase quickly so the wax covered the entire surface before it cooled, and bingo! That old vase was as good as new. (This technique also makes porous surfaces waterproof—which means you can turn just about any old container into a vase.)

Cut-Flower Extender

When I was a boy, I used to pick flowers from Grandma Putt's garden. After she showed me the proper way to pick them (apparently pulling them up bulb and all was not appropriate), I got pretty good at it, and always kept a bouquet in the house for Grandma. To help the flowers last longer, she put this tonic in her vase.

DIRECTIONS: Pour the water into the vase, then add the vinegar and sugar to it. Make sure the sugar dissolves completely, then place your flower bouquet in the mixture. It'll last much longer than in just water alone. And I should know, since Grandma always used to say that I picked the most beautiful blooms!

> 2 tbsp. of white vinegar
> 1 tsp. of sugar
> 1 qt. of warm water

Longer-Lasting Cut Flowers 🏠 ONE OF THE QUICKEST WAYS to make cut flowers stay fresh a little longer is to use hair spray. Just give the posies a light mist of the stuff after they've been in the vase for a day or so. That's it!

Get Corny 🏠 HERE'S A SUPER-SIMPLE SOLUTION for prolonging the life of cut flowers. Fill the vase with a mixture of 2 tablespoons of clear corn syrup per quart of water. The gooey solution will keep your flowers going strong for a few extra days.

Bundle Up 🏠 WANT TO ENJOY FRAGRANT FLOWERS even in winter? After you cut your flowers (and herbs), tie bundles of stems with cotton string, and hang them in a warm, dark, well-ventilated place to dry. Then simply toss a bundle or two in your fireplace for a heavenly reminder of summer.

Super Shortcuts

This recipe won't make fresh flowers last forever, but it'll give you several more days to enjoy their beauty. Just fill their vases with a solution made from 1/8 teaspoon of bleach and 1 teaspoon of sugar per quart of water.

Dynamite Dirt Buster

You might think that bringing a plant inside will keep it safe and healthy, which is certainly true because there are far fewer dangerous pollutants indoors than out. But there are still plenty of things that can hurt your houseplants. Here's a tonic that'll keep your indoor greens happy, healthy, and clean.

DIRECTIONS: Mix all of the ingredients together, and put them in a handheld sprayer bottle. Liberally spray the solution on your potted pals, and wipe off any excess with a clean, dry cloth. Use this mixture whenever you notice your indoor plants looking like they could use a good perk-'em-up.

Lemon Up Container Plants 🏠 NEXT TIME YOU WATER your container

plants—indoors or out—add a few drops of lemon juice to the watering can. It'll lower the water's pH, thereby allowing the plants to take up more nutrients from the soil.

"Babysit" Your Houseplants 🏠 BEFORE YOU TAKE OFF on a trip, line your

bathtub with trash bags, and cover them with a big wet towel. Set your plants on the towel, and just before you leave, water them thoroughly. Assuming the pots have drainage holes in the bottoms, your green pals should stay in fine fettle for two weeks or so.

GRANDMA'S OLD-TIME TIPS

If your indoor plants could do with a little more light, take a tip from Grandma Putt, and set each one on a mirror. Or line a whole windowsill (and maybe even the frame) with mirrors. The sun's rays will bounce off the glass and reflect onto the foliage, giving your green friends the sunshine they crave.

Fabulous Foliage Food

For potted plants that have a whole lot of foliage, offer a taste of this treat to make them flourish. Ferns, philodendrons, and other leafy plants will stay at their lush green best with this elixir.

DIRECTIONS: Mix the ingredients together, and feed your growing greens with this tonic every two weeks or so.

1 crushed multivitamin-plus-iron tablet

½ tbsp. of ammonia

½ tbsp. of bourbon*

½ tbsp. of hydrogen peroxide

¼ tsp. of instant tea granules

1 gal. of warm water

*For flowering houseplants like Christmas cactus, cyclamen, and Persian violets, use vodka instead of bourbon.

Happy, Healthy Houseplants GRANDMA PUTT HAD the happiest, healthiest houseplants. Her secret: She fed each of them once a month with an aspirin tablet dissolved in water. It was just the ticket to keeping her plants green and pretty.

Hold That Soil BEFORE YOU PUT A PLANT in a pot (indoors or out), lay a square of panty hose over the pot's drainage hole to keep the potting mix from leaking out. That way, the plant will be able to maximize its nutrients, and your countertop or windowsill will stay a little cleaner.

Super Shortcuts

Lighten up your potted plants—and improve drainage at the same time. Before you add soil to a pot, spread a layer of foam peanuts on the bottom. How deep should it be? It all depends. For a small pot or hanging basket, an inch or so will do the trick. When you're planting in a large container—especially if it will be resting on a deck or balcony, where weight is a critical factor—fill the pot one-fourth to one-third of the way with peanuts, then pour in a lightweight potting mix.

Flower-Drying Power Powder

> **2 parts cornmeal**
> **1 part borax**

Sometimes you run across a bloom (or a whole bunch of them) that's just too special to get rid of. It may be your wedding bouquet, or it might just be a cute cluster from your perennial garden. Either way, you can preserve your bloomin' beauties with this marvelous mixture.

DIRECTIONS: Mix the ingredients, then sprinkle half of the powder in a plastic shoe box with a tight-fitting lid. Lay the flowers you'd like to dry over the mixture, then lightly sprinkle more powder on top. Seal the container and leave it undisturbed for 10 days. All that's left to do is remove the flowers and gently brush or blow away the powder.

Straighten Up 🏠 I LOVE FLOWERS of all kinds, but some of my favorites for indoor arrangements are big, heavy blooms like tulips, mums, and dahlias. Sometimes, though, a few of the flowers in a vase flop a little too much for my liking. To straighten them up, I stick a straight pin vertically into the stem, just below the bottom of the flower. It works like a charm!

Crazy Daisies 🏠 TURN WHITE FLOWERS BLUE by adding ¼ cup or so of bluing to the water in the vase. The color will travel up through the stems and into the petals. Of course, the more bluing you use, the darker the shade will be. (You can find bluing in the laundry-supply aisle of some grocery stores or online.)

Q *It seems that I don't have the touch when it comes to making my cut flowers look nice in a vase. They end up going every which way, and then drooping. Can you give me some floral design tips?*

A The pros use all kinds of tricks to keep flowers neatly arranged. Here's one to try: Form a piece of chicken wire into a ball and put it in the bottom of the vase (a nontransparent one, of course, so the wire doesn't show through). Then slip the stems right into the openings. With this handy trick, your arrangements will always look lovely.

Hit 'Em Hard Houseplant-Pest Potion

> ¼ cup of gin or vodka
> 2 tbsp. of baby shampoo
> 2 tsp. of hot-pepper sauce
> 1 qt. of warm water

Even when they're inside the comfort of your home, your plants can be attacked by insect pests. No matter what kind of thugs are bothering your indoor beauties, this intensive treatment will send the little buggers to the local pest cemetery.

DIRECTIONS: Mix these ingredients in a handheld sprayer bottle. Then put the plant in your bathtub or kitchen sink, and cover the soil with aluminum foil or plastic wrap to keep it in the pot. Rinse the plant thoroughly with clear water. Then apply this tonic to the point of runoff, being careful to drench all stems and both sides of the leaves, as well as the leaf nodes (the point where a leaf meets the stem—a favorite hiding spot for many small insects). Let the plant sit for 15 to 20 minutes, then rinse thoroughly again with clear water. You'll send your pest problems packing!

Houseplant-Pest Blaster ◆ ARE APHIDS, MEALYBUGS, or other teeny terrors buggin' your leafy pals? Just dab each critter you see with a cotton swab dipped in rubbing alcohol, and you'll kiss those pesky pests good-bye for good!

Cat Deterrent ◆ DOES FLUFFY INSIST on using your potted plants as a litter box? Or maybe she just gets a kick out of digging up the soil. Either way, protect your green friends by laying pieces of fine-mesh screen over the soil. Water and air will get through just fine, but kitty's claws won't.

That's Brilliant!

Want to know a spicy secret for keeping Fluffy and Fido away from your philodendron? Just shove a few cinnamon sticks into the soil in each pot. Cats and dogs both seem to dislike the aroma and will find someplace else to amuse themselves.

Houseplant Perk-Up Potion

Indoor plants can add a little something extra to an otherwise lackluster decor, so they deserve to be kept looking their best. If your plants' appearance is dreary and dull, give them a nutritional boost with this old-time chow.

2 crushed multivitamin-plus-iron tablets
¾ cup of ammonia
1 tbsp. of baking powder
1 tbsp. of Epsom salts
1 tbsp. of saltpeter*
½ tsp. of baby shampoo
½ tsp. of unflavored gelatin
1 gal. of water

DIRECTIONS: Pour the water into a bucket, then mix in the remaining ingredients. Pour the mixture into a container with a tight-fitting lid. (Use several jars if you don't have one that's big enough to hold it all.) Then, once a month, use 1 cup of this potion per gallon of water instead of your regular fertilizer. Your plants will all but jump for joy!

*Available at drugstores.

Stake Out ● IF YOU NEED A STAKE for a floppy young houseplant, grab a pencil. Just shove it into the soil, and tie the plant loosely to the pencil with soft yarn or string. No more floppy ferns!

Container Plant Sticky Trap ● WHEN BAD-GUY BUGS are munching on your potted indoor plants, coat laundry detergent caps with corn syrup, petroleum jelly, or spray adhesive, and set them on the soil in the pot, or nestle them among the branches. One cap will be large enough to debug a small to medium-size plant.

PINCHING PENNIES

If you use talcum powder (or anything that comes in a bottle with a sprinkle top), there's no need to toss the empty containers. Instead, wash them and turn them into mini sprinkling cans for your potted pals or seed-starting flats, or use them to dispense homemade powdered cleansers.

Instant-Aging Formula for Terra-Cotta Pots

> 1 cup of moss
> ½ can of beer
> ½ tsp. of sugar

Some folks love the look of a brand-new beautiful clay planter. Others prefer containers that have a little more experience under their belt—or at least look as though they had. If you fall into the latter camp, then this tip's for you.

DIRECTIONS: Combine the ingredients on low speed in an old blender that you no longer use for food. Then paint the mixture on the outside of your terra-cotta containers. In a week or so, that lovely moss and lichen will start the instant aging process.

Yogurt Terra-Cotta Pots ⌂ TO MAKE NEW TERRA-COTTA pots look as though they've been passed down for generations, just paint them with plain yogurt and set them outdoors in a shady spot. As the yogurt dries, moss and lichen will grow on the clay surface. Keep an eye on the containers, and when they've aged enough to suit you, hose them off with clear water. It generally takes a week or so to produce an authentic-looking "antique."

Disinfect Old Containers ⌂ BEFORE YOU SET a plant in a pot that's been used before, disinfect the container the way Grandma Putt did: Soak it for 15 minutes or so in a solution of 1 part bleach to 8 parts water. Then be sure to rinse it well before you put it back in circulation.

That's Brilliant!

If you're like most folks I know, you've got at least a couple of those giant-size gift tins that once held popcorn or other snack treats. Well, here's how to put one to good use. Spray the inside with a waterproof sealer. Then proceed to either drill five or six ½-inch holes in the bottom of the tin, add potting mix, and tuck in your plants; or pour an inch or two of gravel into the bottom, and set a potted plant on top of the gravel. You've now got yourself a handy planter!

Perfect Pot-Cleaning Potion

Sometimes the insides of my pots develop a slimy film from algae or a white residue from salt leaching out from the soil. Before reusing clay or plastic pots, I wash them first in this easy-to-make solution that prevents slimy and salty buildup.

1 oz. of dishwashing liquid
½ oz. of antiseptic mouthwash
½ oz. of bleach
½ oz. of hydrogen peroxide
1 tbsp. of instant tea granules
1 qt. of warm water

DIRECTIONS: Combine all of the ingredients in a large plastic bucket, put your dingy pots into the bucket, and let them soak for 15 minutes or so. Then take 'em out and scrub 'em clean with a nonabrasive plastic scrubbie. Rinse each pot well, then fill it up!

A Bright Idea 🏠 WHEN YOU BRING YOUR HERBS and geraniums indoors for the winter, set the pots on windowsills that you've lined with aluminum foil, shiny side up. It'll reflect light onto the plants and keep them going strong all winter long.

Winterize Outdoor Container Plants 🏠 IF YOU CAN'T MOVE your large pots indoors for the winter, protect the plants' roots by covering each container with two or three layers of bubble wrap, held in place with twine or duct tape. This overcoat offers as much insulation as a couple of feet of snow would provide for in-ground plants.

Super Shortcuts

✂ Those lush, beautiful houseplants sure do look nice—but they're real dust catchers. So when it's time to dust your furniture, take a minute to dust your foliage, too. If the leaf stems break easily, hold them underneath for support before you send the dust flying. You can use a feather duster, a synthetic duster, or just an old T-shirt to do the job. For houseplants with fuzzy or hairy leaves (like African violets), use a small soft paintbrush to flick the dust off.

Root, Root, Root for Houseplants Potion

One of my favorite ways of starting new house-plants is by taking stem cuttings of plants I already love. The resulting rooted plants are identical to the plants I collected them from, and they grow lush and lovely in the blink of an eye. Be sure to take your cuttings from plants that are healthy and mature. But don't place the tender stems in plain old water. Give it a boost with this peppy potion.

DIRECTIONS: Pour the water into a clear vase or drinking glass. Then stir in the soda and ammonia, and add your cuttings. Once you see that the roots are off and running, pot the youngsters up in small containers, using a light, well-drained potting mix. When the plants are well established, move them into their permanent dwellings.

Blushing Beauty ⌂ THE CHUBBY BRUSH that comes with powdered blush or bronzer works great for cleaning fuzzy-leaved African violets, spiny cacti, or sharp-edged pineapple plants. It has just the right amount of stiffness to coax the dust out of the crevices, and it'll keep your fingers out of harm's way, too! You can also use it on the clustered leaves of bromeliads and on other plants that have lots of nooks and crannies.

GRANDMA'S OLD-TIME TIPS

Whenever I finished eating a banana at my Grandma Putt's house, I knew better than to throw away the peel. Why? Because a banana peel was Grandma's favorite tool for polishing her big philodendron leaves. To use her secret method on any houseplant with large smooth leaves, simply hold each leaf of the houseplant and wipe the surface with the inside of the peel. The natural oil in the banana skin gives the leaves a nice shine.

Smooth Leaf Shiner-Upper

½ cup of nonfat dry milk
1 qt. of warm water

Like everything else that sits stationary in your house, leafy greens can collect dust. Just use this simple solution to shine the leaves of your ivy, philodendron, or other smooth-leaved plants. It'll wipe away the dust and leave your leaves gleaming.

DIRECTIONS: Mix the milk and water in a bucket, dip in a cellulose sponge, and use it to wipe the top and bottom of each leaf. Wipe gently and don't rub, or you'll risk damaging the natural coating on the leaves. When you're done, there's no need to rinse; just let the milky mixture dry to a shine.

Hair Spray Cleanup 🏠 WHEN APHIDS START PLAGUING your houseplants, reach for the aerosol hair spray and a plastic bag that's big enough to hold the plant and its pot. Spray the inside of the bag—not the plant! Then put the victimized plant inside, fasten the bag tightly with a twist tie, and set it in a spot away from direct sun (otherwise, the heat inside will build up and kill the plant). Wait 24 hours, and remove the bag. Those itty-bitty bad guys will be history! **Note:** This trick works on outdoor container plants, too.

Off to a Good Start 🏠 HERE'S A DANDY WAY to make your own seed-starting "pots." Collect as many laundry detergent caps as you'll need, punch or drill drainage holes in the bottom of each one, and set them in a tray or shallow pan. Then fill each cap with starter mix to within about ½ inch of the top, and plant your seeds.

Q *My indoor plants are looking a little dull. Even after I dust them, they still look a bit dingy. Is there a safe way that I can get them shining bright again?*

A "Polish" your houseplants the old-fashioned way with citrus juice (any kind will do). Just wipe the juice onto the leaves with a soft, clean cotton cloth. It'll leave the foliage sparkling clean—and better able to absorb carbon dioxide from the air.

Tea-rific Treat

Our potted pals purify the air we breathe, add color and interest to any room, and make a house feel more like a home. I can't imagine living without them! So every now and then, I whip up this top-notch tonic that keeps all of my houseplants in peak condition.

> 1 used tea bag
> 1 qt. of warm water
> 2 drops of ammonia
> 2 drops of antiseptic mouthwash
> 1 drop of dishwashing liquid

DIRECTIONS: Pop the used tea bag into the water, and add the ammonia, mouthwash, and dishwashing liquid. Then pour this energized "tea" into a handheld sprayer bottle, and spray your plants to clean them up and protect them from disease.

Keep 'Em Quenched, Not Drenched 🏠 IF YOU DON'T HAVE the patience to go through your house every few days, sticking your finger into the soil of your plants to see if they need watering, I've got the answer for you—in a word, pinecones! Yep, these are Mother Nature's own moisture meters. Just stick a fresh pinecone about two "petals" deep into the soil of each of your potted plants. The petals of the cone will open when the soil is dry and close when it's wet. So there's no more wondering when it's time to water!

PINCHING PENNIES

When it comes to watering plants, there's good water and bad water. The best water? Rain or melted snow, which, of course, is free. So take advantage of precipitation whenever you can. The worst water for your plants is the stuff you pay your municipality for that comes straight out of the tap. The good news is that you don't need to spend a small fortune on bottled water or fancy filters. You can turn ordinary tap water into plant-friendly water by adding a layer of agricultural charcoal to the top of the soil in your plants' pots. It'll filter out additives like chlorine and fluoride, making the water perfect for your plants. And as a bonus, the charcoal will keep your house smelling fresh. (You can find agricultural charcoal at garden-supply centers.)

Turbo-Boosted Wonder Water

There's no big secret to keeping houseplants happy and healthy. Just keep their pots clean, give 'em fresh potting soil every so often, keep 'em well fed and watered, and take care of any pest problems as soon as you spot them. To make this TLC even more tender-lovin', water your plants with this healthy H_2O.

> 1 tbsp. of Epsom salts
> 1 tsp. of baking powder
> ½ tsp. of ammonia
> 1 gal. of water

DIRECTIONS: Mix all of the ingredients in a bucket, then transfer the solution to a watering can. Give your green friends a healthy sip of this water once a month; stick to your regular feeding and watering regimen the rest of the time.

Sock It to Me! 🏠 GIVE PALM TREES and other smooth-leaved plants the white-glove treatment by putting a pair of soft, white cotton socks on your hands when you clean them. Wet the socks under lukewarm water, make a fist to squeeze out the extra moisture, and then slide each leaf between your, ahem, palms. Use only very gentle pressure so you don't wipe off the natural waxy coating on the leaves.

Make It Shine 🏠 IF YOU LIKE A HIGH-GLOSS look on your houseplants, check your fridge for these shiner-uppers: Use a bit of mayonnaise on a soft, clean cloth to gently wipe the leaves. Or, beat an egg white to a froth and wipe the foam on the leaves. Either one of these tricks will give your plants a great sheen.

That's Brilliant!

When it's feeding time for your ferns, ficus, and philodendrons, and you're fresh out of their usual chow, just give them a drink of beer instead. The brew is one of the best all-purpose fertilizers a plant could ask for, and it's completely safe for them to drink!

Winter Wonder Drug

It's always nice to have some of your foliaged friends in your house during the winter months. They help remind us that winter is only temporary, and soon enough the outdoors will be lush and green again. To keep indoor herbs, veggies, and citrus trees happy and healthy all winter long, douse them with this terrific tonic every now and again.

1 tbsp. of liquid kelp
½ tsp. of dishwashing liquid
1 gal. of water

DIRECTIONS: Combine the ingredients in a bucket, then pour some of the mixture into a handheld sprayer bottle and mist-spray your plants every two to three weeks. It'll feed 'em and fight pests in one easy step, and your indoor plants will stay their happiest and healthiest all winter long.

Perk Up Ferns 🔺 A DOSE OF CASTOR OIL will save a fern that's looking less than fabulous, so try this trick on your ailing plants. Combine 1 tablespoon of castor oil and 1 tablespoon of baby shampoo in 1 quart of warm water, and give each plant ¼ cup of the solution. Your ferns will turn green and fresh again almost overnight.

Keep 'Em Colorful 🔺 MAINTAIN THE VIVID COLOR of your houseplants by mist-spraying the foliage every week or so with a mixture of 1 twice-used tea bag, 3 teaspoons of ammonia, 3 teaspoons of mild shampoo, 1 teaspoon of antiseptic mouthwash, and 1 quart of warm water. This gentle shower will keep the bright green of the leaves, or other vibrant colors, sticking around long after they would naturally fade.

Super Shortcuts

✂ If you water your plants with tap water that has a lot of dissolved minerals in it, or if you mist-spray your plants with fertilizer, the leaves can start to look dull and drab. What to do? Wipe a little pineapple juice onto the greens with a soft, clean cloth. They'll shine right up!

CHAPTER 13

OUTDOOR
Chores

When it's time to roll up your sleeves and tackle outdoor chores, you don't have to spend a lot of money and labor to get the job done. In this chapter, you'll find all the tools you'll need in the form of my time-tested advice. Whether you're scrubbing your house siding, sprucing up your deck, washing your car, or removing rust from garden tools, my tonics and tricks will help get the job done in no time flat. Then you can sit back, relax, and enjoy the great outdoors!

All-Surface Cleanup Solution

Are you looking for a cleaner that will make everything on the outside of your house spic-and-span? Well then, here's a handy recipe that you can use on stucco, concrete, or any other dirty outdoor surface.

> ⅔ cup of powdered household detergent like Spic and Span®
> ⅓ cup of powdered laundry detergent
> 1 gal. of water

DIRECTIONS: Mix all of the ingredients together in a bucket, put on a pair of rubber gloves, grab a sponge, and you're ready to rumble. If you need to scrub stubborn stains, include a stiff-bristled brush in your arsenal. And if you're cleaning a surface that's mildewy, use less water and add 2 cups of household bleach to the mixture.

Can It! 🏠 HERE'S A WAY to decoratively light up an outdoor area like a patio or porch with candlelight. All you need is an aluminum can, a nail, a hammer, paper, a pencil, and some masking tape. The idea is to punch holes in the can to let the candle-light shine through. But hammering on an empty can will only flatten it out—unless you know this trick. Start by washing and drying the can, and removing the label. Fill the can with water and place it in the freezer until it's frozen solid. The ice that fills the can will help it retain its shape while you pound away at it. Once the water is frozen, use the pencil and paper to work up a design for the holes. Then stick this template to the side of the can with the tape. Using the nail and hammer, drive small holes along the lines of the design and through the can walls. When you've finished punching out the design, let the ice melt and you're ready to put a candle inside.

PINCHING PENNIES

Are you finding rust on your wrought-iron patio furniture? Not to worry. Just rub those rusty spots with kerosene, and lightly scour them with very fine steel wool. If the rust is stubborn, let a kerosene-soaked rag sit against it for a while. Then, to keep that rust from making a comeback, give the furniture a good going-over with a dollop of car wax. Yep, car wax. It's designed to protect the finish of cars and works just as well on all kinds of patio furniture, too!

Automotive Anti-Frost Formula

> 6 cups of white vinegar
> 2 cups of water

There are few more annoying wintertime woes than waking up to find your car windows covered in frost or—worse—a thick coat of ice. This ultra-simple elixir can end your ice-scraping days for good (provided, that is, you remember to use it every time you have to leave your car parked out-doors on a frosty night).

DIRECTIONS: Mix the vinegar and water in a bucket or pan, drench a cloth with the solution, and give all the windows and mirrors a good once-over. (Or, if you'd prefer, pour the solution into a handheld sprayer bottle and spritz it all over the glass.) The next morning, you'll wake up to frost- and ice-free windows.

De-Stick Decals ⌂ OLD CAR WINDOW DECALS and stickers can be tough to remove, but fortunately, the glass they're stuck to can withstand harsher treatment than the painted or chrome finish of a bumper. Soak the decal with lighter fluid or nail polish remover for a few minutes, then carefully scrape it away with a razor blade. **Note:** Lighter fluid and nail polish remover are highly flammable. So don't use them around a hot engine or on a car that's been sitting in the sun.

Keep a Level Head ⌂ WHEN YOU'VE GOT a flat tire and are looking for a safe place to pull off the road, try to find an area that is (1) level and on solid ground, (2) away from traffic and curves, and (3) in plain sight of other drivers. However, don't spend too much time searching for a perfect spot—driving on a flat can damage the tire's rim and possibly other parts of your car. When you do find a good spot, be sure to apply the hand brake to prevent the car from rolling away.

That's Brilliant!

Here's a handy item to have when your car windows steam up on the inside. (It's a trick I picked up from an old friend who was a teacher, and it's a lesson I'm glad I learned!) Carry a felt blackboard eraser in your glove compartment for whenever things get steamy. A couple of wipes will quickly remove the condensation.

Car-Savin' Salt Remover

1 cup of dishwashing liquid
1 cup of kerosene
Water

If you live in snowy-winter territory, you know that the road salt can build up on your car and shorten its life span considerably. So keep this simple recipe handy, and use it to give your car a good going over every time you come home from driving on wet, salt-treated roads.

DIRECTIONS: Mix the dishwashing liquid and kerosene in a bucket of water, and use a sponge to rub the salt deposits away. (But remember: this stuff is flammable, so use it with caution!) Don't pour it out on your driveway, either, because then you'll just have to use another tonic to wash down the drive.

Plan Ahead 🏠 ARE YOU ONE OF THOSE folks who jumps into the car every single time you need something from the store? If so, shame on you! One of easiest ways to save gas is to do a little planning. At the beginning of the week, sit down with your family and make a list of all the errands that need to be done. Organize errands by location, and then tackle them all in one day. Not only will this little trick save you gas money, but it will save you time, too. Of course, if you're within walking or pedaling distance of any of your destinations, leave the car at home.

PINCHING PENNIES

Keeping your tires properly inflated will help your car run more smoothly and keep the wheels in alignment. Here's how to tell whether your tires are under- or overinflated, and it won't cost you a nickel, either, although you will need a quarter. For each tire, insert the quarter into the tread at the inside, center, and outside of the tire. If the quarter fits deeper into the treads at the edges than in the center, the tire is overinflated. If the tread is deeper in the center than at the edges, then the tire needs more air.

Clean as a Whistle Hand Cleanser

When you spend as much time puttering in the yard and garden as I do, your hands get mighty dirty. I get mine as clean as a whistle—fast—with this superpowered cleanser. Keep a container of it beside each outdoor faucet, another in your garden shed, and another in the garage, so it's always close at hand. (Pun intended.)

DIRECTIONS: Pour the water into a large bucket or tub, add the borax, laundry detergent, and cornmeal, and mix to form a paste. Spoon the paste into lidded containers (sturdy plastic food-storage containers are perfect), and label them clearly. Then stash one in each place where you tend to get your paws grubby.

Wash-and-Wear Garden Gloves 🏠 NOW HERE'S A NEAT TRICK—clean your garden gloves while you are still wearing them! Just lather them up with soap and water after you finish working, and scrub 'em together to get rid of the grime. Rinse them under running water, strip them off, roll them in a towel, and give them a squeeze to soak up the water. If the gloves are leather or suede, lay them flat to dry; if they're cloth, just clip them to your clothesline. They'll be stiff as a board when they're dry, but don't worry—they'll soften up as soon as you start digging in to the next round of garden work.

Super Shortcuts

✂ For quick and easy cleanup once you're done working in your flower garden, place a bar of soap in the toe of an old nylon stocking or panty hose leg, and hang it near an outdoor faucet. Instead of making a mess inside the house, you can scrub up outside without even having to take the soap out of the stocking. Believe you me, this simple secret has saved many a gardener's marriage!

Crackerjack Concrete Cleaner

Use this mixture to scrub away stains and brighten the appearance of your sidewalks, driveways, patios, or outdoor walls. You can always do the scrubbing by hand, but it's a whole lot easier to let a power washer supply the muscle.

DIRECTIONS: Mix this potion outside, so you don't breathe in the fumes from the bleach, and wear rubber gloves, goggles, long sleeves, and old pants when you use it to guard against stray splashes. Then cover all surrounding vegetation, buildings, and materials with plastic sheeting, so they don't get damaged by any overspray. If you're working by hand, wet down the area with a garden hose, spread the bleach solution on with a long-handled scrub brush, and use lots of elbow grease to get rid of those stains. Rinse the area thoroughly after scrubbing. If you've got a power washer standing by, spread the solution over the wet concrete, and let the blast of water do the scrubbing for you! Then rinse the area, and pack up the power washer till next time.

Rub Out Rust ⬆ WHEN RUST STAINS MAR your outdoor concrete, buy a bottle of rust remover that contains oxalic acid to bleach out the stain. Apply the rust remover and let it soak in for about an hour, then rinse the area clean with a hose. Be sure to protect your eyes and skin with goggles and gloves (and all surrounding vegetation, buildings, and materials with plastic sheeting) when you work with this compound, or any other cleaner that contains acid.

That's Brilliant!

If splashes and drips from your hot tub have left white stains on the concrete around it, blast them away with a power washer. The stains are just mineral deposits from the chemically treated water, and a powerful blast of good ol' H_2O is all you really need to make them disappear in a flash.

Down-Home Deck and Porch Perfectionizer

> 1 qt. of bleach
> ½ cup of powdered laundry detergent
> 2 gal. of hot water

Grandma Putt's house had big, wide wraparound porches, and she spruced them up every spring and kept them spotless with this simple formula. (It works just as well on any wooden structure, including 21st-century decks.)

DIRECTIONS: Mix all of the ingredients in a bucket, and scrub the porch or deck using a stiff broom or brush. Then hose it down thoroughly. You'll wash away that winter gunk and anything else Mother Nature left behind.

Take the Gray Away

IF YOUR DECK is looking rather dreary because sunlight has faded the finish to a dull gray, try an oxalic acid deck cleaner to restore the natural beauty of the wood. Read the label before you begin, and follow the directions carefully. You'll probably need to apply the cleaner with a rag one board at a time, scrub it in with a soft brush, and rinse thoroughly with water. And don't forget—acid is mighty powerful stuff, so dress in protective clothing, wear rubber gloves and goggles, and cover all surrounding vegetation, buildings, and materials with plastic sheeting, so they don't get damaged by any overspray.

PINCHING PENNIES

There's no question about it: A covered, enclosed porch is a big investment. So before adding one on, you must ask yourself if it's worth the price. The answer is probably yes if you share your backyard with a lot of hungry insects. And if you want to use the space nearly year-round, and you don't live in a balmy climate, it's a less expensive option than adding a room on to your house. Plus, you could always use the extra space for visiting friends or family members to sleep in.

Dynamite Driveway Cleaner

If you've got oil stains on a concrete driveway or garage floor, you may be tempted to scrub them away by hand. But save yourself—and your back—the trouble, and turn to this concrete-cleaning solution instead.

> Paint thinner
> Clay cat litter
> Stiff-bristled
> push broom

DIRECTIONS: Pour some of the paint thinner onto any visible spot on the concrete, covering an area at least 6 to 12 inches beyond the stain. Then spread the cat litter thickly over the entire treated area, making sure that none of the treated concrete is visible. Let the cat litter stand for about an hour to absorb the grease. Then use the broom to sweep up the saturated cat litter. Repeat the process as needed until the surface looks (almost) as good as new!

Clean Up Oil

GRANDPA PUTT LOVED to putter with his car, and he was mighty careful about keeping oil from spilling onto the garage floor. Once in a while, though, even he would end up with messy splotches on the concrete. He cleaned them by covering the spills with equal parts of baking soda and cornmeal. He'd wait until the oil had been absorbed, and then he'd sweep up the residue. If any stains still showed, he'd wet the floor with clear water, and scrub the spots with baking soda and a stiff brush, and rinse.

FAQ?

Q *I've got a couple of concrete blocks left over from a project. My wife is bugging me to get rid of them, but I'm sure there is something I can do to put them to good use. Can you back me up on this?*

A I sure can! One clever thing you can do with concrete blocks is to make a portable step. Just wrap a block in thick foam rubber or felt, and cover it with sturdy fabric or textured wallpaper (to provide traction). Then set it wherever you need a little height boost—for instance, in front of the sink in a child's bathroom, or in a closet with a shelf that's just out of reach.

Flawlessly Fabulous Outdoor Window Wipe

The outsides of your windows can get just as dirty (probably even more so!) as the insides. Here's a terrific solution to cleaning the outsides of your windows without leaving any streaks behind—and without using a squeegee.

> ½ cup of liquid dishwasher detergent
> 2 tbsp. of dishwasher rinsing agent
> 2 gal. of hot water

DIRECTIONS: Combine the ingredients in a bucket, then scrub the mixture onto your windows with a brush, and rinse them clean with a garden hose. The water will roll off the windows, leaving nothing behind but a beautiful shine.

Scratch the Scratch 🏠 NOW THIS ONE'S GOING to surprise you. How can you get small scratches out of glass without some serious polishing? Simple—you scratch it back! Rub a little toothpaste (the white kind, not gel) into the scratches with a toothpick. Then polish with a soft cloth. Voilà! No more scratches!

Window Wonders 🏠 THERE ARE SO MANY WAYS to clean windows, it's hard to know which method to use. Try lemoning up next time you tackle the chore. Combine 1 tablespoon of lemon juice with 3 cups of water. Apply the solution to the glass with a soft cloth and dry the surface with a second soft cloth. Your windows will sparkle, and nothing beats that fresh lemon scent. Or, if you're not fond of scented windows, clean them with a little club soda on a soft cloth. Dry immediately with a second soft cloth for a picture-perfect view.

Super Shortcuts

✂ A lot of folks say "I don't do windows" because it's so frustrating to get out there and wash and rub, only to see a ton of streaks when the sun shines in. Well, my solution to the streak problem is simple—vinegar! It's great for cleaning glass. Just combine equal parts of white vinegar and water, and apply the mixture to your windows with a soft cloth. Dry with a second soft cloth to leave your windows nice and sparkly—and virtually streak-free.

Good Riddance to Bad Sap Solution

Laundry detergent
Very hot water
Auto body polish

If you park your car outside every night, you might notice some gooey buildup on its surface in the morning. Tree sap starts flowing freely in warm weather and (ugh!) hardens up once it gets on your car. This formula will take that hard goo right off without damaging the finish.

DIRECTIONS: Mix the detergent and water together in a bucket to get a lot of suds started. Take a spatula and press down on a globule of sap to break its bonds. Then dip a cloth into the water, picking up a lot of suds. Place it on top of the glob of sap and press down as hard as you can. Repeat this process until the glob disappears. If the water gets cool, make up another batch with hot water. When the sap is gone, spray down the areas with cold water, then spiff them up with the auto body polish as directed on the label.

Quick Sap Fix 🏠 THE NEXT TIME you find sticky sap on your car, you need to act fast. If you catch it before it hardens, it's easy enough to get rid of. Just grab a bottle of rubbing alcohol and a clean, soft cloth. Then dampen the cloth with the rubbing alcohol and rub out the sap spots. Finish up by rinsing the areas with cold water and wiping them dry with a clean towel.

That's Brilliant!

There's a simple way to remove dried car wax from your black plastic bumper strips. Just whip up a mixture of $1/2$ cup of baking soda in 1 cup of turpentine or mineral spirits. Apply the mixture to the entire bumper strip, then scrub it with a brush. Rinse the area well, and you should be good to go. Spots will generally disappear with one or two rounds of scrubbing.

Grandma Putt's Baby-Rockin' Wicker Preservative

Grandma rocked all her babies, grandbabies, and even great-grandbabies to sleep in a big, old wicker rocking chair. She used this formula once a year to keep it—and all her other wicker furniture—in tip-top condition. It worked, too. I know that for a fact because my wife and I have rocked our own babies and grandbabies to sleep in that very same chair!

DIRECTIONS: First, dust the piece with a soft brush. Then mix the linseed oil and turpentine in a wide-mouthed glass jar. Rub the solution into the wicker with a soft cloth, paying special attention to all the nooks and crannies. Remove any excess formula with a clean, dry cloth, and let the furniture air-dry.

Keep It Under Cover 🏠 WICKER FURNITURE CAN'T STAND UP to the elements the way teak, redwood, and metal can. Even in its ideal summer home—a covered veranda—wicker needs more attention than tougher materials. So if you know that a storm is brewing, bring your wicker furniture indoors to avoid any damage.

GRANDMA'S OLD-TIME TIPS

Every time I see an old-fashioned porch all decked out with wicker tables, chairs, and rockers, it takes me right back to my younger days with Grandma Putt. When it came to caring for it, she was as fussy as a mother hen with her chicks. And, as always, she bestowed a lot of her know-how on yours truly. Once a year (or, more often, if it's exposed to dust and grit), vacuum your wicker furniture with the machine's soft brush attachment, or brush it by hand with a soft brush. To keep white wicker bright without painting, mix ¼ cup of salt in 1 gallon of warm water. Dip a stiff brush into the warm salt water and give the furniture a good scrubbing. Then let it dry in the sun.

Headlight Sparkle Solution

2 tbsp. of ammonia
1 tbsp. of cornstarch
1 qt. of water

Car headlights can get really grimy, especially in the winter. After all, those front-end lamps lead the charge through slushy, salty streets. Cut right through that road crud with this super solution.

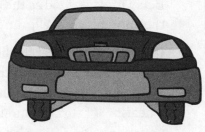

DIRECTIONS: Mix the ingredients together, then wipe the solution onto the lights with a soft, clean cloth, and rinse with clear water. The cornstarch will act as a mild abrasive, and you'll be seeing your way clear in no time.

Be Prepared 🏠 BEFORE YOU SET OUT on a trip—especially in winter—make sure that all of your car's components are in good working order: your battery, brakes, defroster, heater, lights, and windshield wipers. And check all fluids, too: antifreeze, oil, and windshield-wiper solvent. Once you're satisfied that the fluids are at their recommended levels, you're ready to roll.

Give Bugs a Wash 🏠 I'VE HAD GOOD LUCK removing stuck-on dead bugs from my headlights with a mixture of 3 parts glass cleaner (the blue kind) to 1 part dishwashing liquid. Let the mixture sit for a minute, and then gently wipe off the bugs with a damp sponge. If there's still some sticky residue left behind, tackle it with a nonabrasive plastic scrubbie that's been dipped In the solution.

Wash Those Wipers 🏠 THERE'S NO POINT in cleaning your windshield if you don't take care of the wipers, too. To remove the grease and grime that accumulate on wipers, use a little baking soda on a damp sponge. It'll take the gunk right off.

Super Shortcuts

Baby wipes are handy to keep around even when you don't have a youngster living with you. In fact, these premoistened disposable towels are just the ticket when you need to clean off your grimy headlights in a flash. It'll do the trick lickety-split—and won't risk damaging any surrounding trim.

Homemade Car Wax

The next time you give the old jalopy a good top-to-bottom cleaning, follow up with a coat of this wonderful whiz of a wax. It'll make Ol' Betsy look as good as the day she rolled out of the showroom. Well, almost as good.

2 cups of linseed oil
2 tbsp. of beeswax*
2 tsp. of carnauba wax
½ cup of white vinegar

DIRECTIONS: Heat the linseed oil and waxes in a double boiler until the waxes have melted. Stir to blend the ingredients. Remove the pan from the heat, stir well, and pour the solution into a metal container. Set it on a heat-proof surface, and let the mixture cool. Apply it to your car using a soft, clean, lint-free cloth. When the wax has dried, very lightly dampen a second soft, clean cloth with the vinegar and gently wipe the car's surface. Then stand back and admire your lovely reflection in the hood!

*Available at craft-supply stores.

Wax Your Shovels WHEN THE SNOW STARTS to fall, grab the car wax. Apply some to your snow shovel blade, and then when you're out shoveling the driveway, the snow won't stick. You'll save lots of time and energy so you can get back inside where it's toasty warm a whole lot sooner!

Q *I used to love finding witty bumper stickers in college and putting them on my car. I thought "Don't Mess with Texas" was the funniest thing ever! But now I'd feel better if I could just get rid of them. Is there an easy way to remove stuck-on stickers?*

FAQ ?

A The short answer is no. You're just going to have to put some real elbow grease into scraping those sticky things off. But for the future, affix bumper stickers only after you've applied a thin layer of car wax to the bumper. That'll make them easier to remove later. It also helps if you strip them off within four to six weeks of application.

Leather-Seat Revitalizer

> ¼ cup of wheat germ oil
> ⅛ cup of sweet almond oil
> 2 tbsp. of castor oil

On a long drive—or even a short one—nothing feels as comfortable on the posterior as leather-covered seats. But if you're lucky enough to have a car that sports these top-of-the-line upgrades, you know that over time, leather can dry out, start to crack, and lose the suppleness that makes it so darn comfy. Well, don't fret. This quick fixer will get your seats back in shape in no time.

DIRECTIONS: Pour the oils into a jar and shake well until they're thoroughly blended. Apply the solution to the leather using a soft, clean cloth (cotton flannel is perfect). Wait one hour, then buff using a second clean, dry cloth. If you use this formula regularly, your luxurious leather seats will stay soft and pliable for many years to come.

Steam Out Smoke Smells

WHEN FANCY LEATHER CAR SEATS smell like stale smoke, freshen them up with a citrus-scented steam bath. Pour about 2 cups of boiling water over the sliced peel of a grapefruit or a couple of oranges. Strain the liquid into a large bowl. Fill the bowl with a few inches of very hot water, and place it in your car with the doors closed for a couple of hours. Presto—no more smoky smell!

GRANDMA'S OLD-TIME TIPS

Petroleum jelly was one of my Grandma Putt's favorite handy helpers—she used it to soothe chapped lips, moisturize dry skin, and even polish shoes. Its oil base makes leather items supple and shiny, so just dab a bit on with a soft cloth, and rub it vigorously to make the leather shine. What kind of a cloth works best? Why, my Grandma had another magic trick up her leg—a clean pair of old panty hose! They make a perfect polishing cloth for a petroleum jelly shine.

Longer-Life Battery Bath

½ box of baking soda
2 qts. of cold water

A good car battery should last almost as long as your car does. And you can keep it humming right along if you make caring for it part of your routine maintenance. Your old jalopy will keep on ticking if you clean your battery twice a year with this solution.

DIRECTIONS: Protect yourself before starting by putting on rubber gloves, a rubber apron, and safety glasses. Disconnect the cable connectors from the terminals. Mix together the baking soda and water in a bucket, and scrub the top of the battery with a scrub brush. Then scrub the connectors, the terminals, and the sides you can reach. If acid residue is present, it'll make a sizzling sound as you clean it. Rinse the battery with plain water. Repeat the process until the sizzling stops, and then dry everything off with a clean rag.

Petroleum Jelly Protection 🏠 IT NEVER FAILS: When your car battery goes kaput, it's always on the coldest day of the year. Low temps make the battery work extra hard, and if it's also crusted with crud and corrosion, well, who can blame the thing for giving up? So after cleaning the terminals, coat them with petroleum jelly to keep corrosion at bay.

Put a Notch in Your Key 🏠 THE CAR, THE HOUSE, or the office? Which key goes where? Here's how I keep them straight. I put a notch in the head (not the jagged part) of my car key with a metal file. That way, I can get into my car quickly, and get on with the business at hand.

That's Brilliant!

During the winter months, put a drop or two of synthetic oil on your car key and slide it in and out of the door locks a few times. The oil will coat the inside of the locks, which will prevent them from icing up in bad weather. But don't try this with regular motor oil or household oil because it will thicken up in the cold and gum up the lock.

Marvelous Molasses Rust Remover

Believe it or not, molasses works just as well as any commercial rust cleaner. Use this sticky solution to make rusty tools, car parts, pots and pans, or other metal objects look like new again.

DIRECTIONS: Getting the molasses out of the bottle is the hardest part of this trick. To make the thick syrup pour more easily, zap it in the microwave in 15-second increments, checking it after each cycle. Once you've got the molasses moving, stir it into the water in a plastic bucket, add your rusty objects, and let them soak. Check the object every day or so to see if the molasses has worked its magic. It may take three or four days before the rust gives up its grip, but it will eventually rinse right off. When the treatment is complete, rinse the objects thoroughly to remove any sticky residue.

Steel Yourself 🔺 TO CLEAN RUSTY hand tools, scrub the problem areas with steel wool. Wet the tools first, then rub the rust away with small circular motions. Rinse the residue off, dry the tools, and coat them with a little mineral oil or rust inhibitor to prevent the rust from creeping back in.

Milk Does Your Tools Good 🔺 AN OLD METAL milk-delivery box—the kind that the milkman used to leave home-delivered milk bottles in—makes a great weatherproof storage bin for handheld garden tools. If you don't have one, look for these old-time treasures at garage sales, or in the want ads.

Super Shortcuts

✂ When rust sets in on hand tools, try this quick fix: Spray the metal with WD-40®, and rub it lightly with a nonabrasive plastic scrubbie. The rust will come right off, and you won't scratch the tools. Finish the job by wiping off the excess lubricant with a dry cloth, and store the tools in a clean, dry place.

Mighty Mold and Moss Remover

Moss, algae, and black mildew can make your patio look like the dickens. What's worse, they also make the surface so slippery that it's an accident waiting to happen. Whichever kind of slimy stuff has invaded your place, this fabulous fixer will send it packing and discourage it from coming back, at least for a while.

1 cup of trisodium phosphate (TSP)
1 gal. of oxygen bleach
1 gal. of hot water

DIRECTIONS: First, dress in protective clothing, pull on a pair of rubber gloves and goggles, and cover all surrounding vegetation, buildings, and materials with plastic sheeting so they don't get damaged by overspray. Then wet down your patio with a garden hose, and mix the ingredients in a bucket. Dip a scrub brush into the solution, and start scrubbing. Let the solution sit for about 15 minutes, and then tackle the toughest stains one more time before thoroughly rinsing everything with the garden hose.

Deck-Mildew Remover ⌂ DAMPNESS CAN CAUSE MILDEW and algae to make themselves right at home on your deck, so use this super solution to make them vanish. Just mix 1 part water with 2 parts bleach, pour the liquid onto the slimy spots, and let it sit for just a minute or two to kill the mold or algae. To avoid leaving unsightly lighter spots behind, use an oxygen bleach instead of a chlorine product. Scrub the stains with a long-handled brush, and rinse the entire deck with a hose to make sure all of the bleach is safely washed away.

That's Brilliant!

You keep every brick, stone, and concrete walkway at your house spic-and-span, just like the rest of the house. But then, despite your best efforts, moss starts to creep in. If you don't care for the mossy look, you can banish it in a flash. Just spray that green fuzz with a solution of 2 tablespoons of rubbing alcohol per pint of water, then rinse it away with a garden hose.

No More Hazy Film Formula

> 1 oz. of liquid dishwasher detergent
>
> 2 qts. of warm water
>
> 4 oz. of ammonia
>
> 1 gal. of cold water

If your car has a vinyl interior, then you're going to love this recipe. Vinyl contains a softener (called a plasticizer) that is used to keep the material pliable no matter how many years you have your car. Pretty great, right? The problem with this plasticizer is that it's given off as a gas when the vinyl gets hot. The gas will settle on the interior windows of the car and produce that hazy film, which I'm sure you're familiar with. This recipe will give you a great haze buster for just such a scenario.

DIRECTIONS: First, mix together the detergent and warm water. Use a rag to wash all the vinyl surfaces in your car—seats, door panels, sun visors, and the padded dash. This mixture will wash away any excess plasticizer that might be sitting there. Wipe it all dry with a clean rag. When that's done, mix together the ammonia and cold water and wash the windows with a sponge. Dry them with paper towels. Depending on the average temperature in your area, the solution may last up to three months (before more plasticizer is released).

Prevent Battery Drain 🏠 IF YOU NEED TO LEAVE your car doors open for an extended period of time (for example, to air it out after you wash your windows with ammonia), the interior light will stay on, draining the battery. To prevent this, push the switches into the doorjamb of each open door and tape them in place with electrical tape. With this simple trick, the lights will stay off, even while the doors are open.

Q *I have an older car (but not too old!) and the antenna automatically slides up and down when the car is on. But lately it's been sticking on the way up. Is there any way to prevent this problem?*

A Try using an old candle stub or piece of sealing wax to occasionally lubricate the antenna while it's in the fully extended position. Waxed paper will do the trick, too, giving you full-time reception of all your radio stations.

Quick and Easy Bicycle Shine

When I was a youngster, if Grandma thought I could do something on my own, she made darn sure I did it. She wanted me to know how to take care of myself if she wasn't around. One of my first responsibilities was to keep my bike looking its absolute best. So every weekend in the spring and summer, I would get out and polish up my bike. Here's a quick and easy way for kids to make their bikes nice and shiny, the same way I was taught.

DIRECTIONS: Fold up each sheet of wax paper into a pad. Hold the pad in your palm, and rub it back and forth along the bike's frame, fenders, and handlebars (but not the handgrips or pedals). When the wax wears off the surface of the paper, flip it over and use the other side. Finally, polish the waxed parts of the bike with the soft cloth until they shine like the sun.

Recycle the Wagons 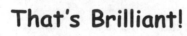 AH, THE OLD RED Radio Flyer® wagons. I loved mine when I was a kid, and they're still pretty popular today. Once your kids and grandkids outgrow their wagon, put it to use in the garden hauling heavy bags of fertilizer and seedling flats. Or, if the body of the wagon is in such bad shape that the only thing it's good for is the scrap heap, remove the wheels and use them as replacement wheels on your lawn mower. They work just as well as the originals.

That's Brilliant!

Before you take off on a long bicycle ride, spray your bike from stem to stern (except for the seat, pedals, and handlebars, of course!) with lubricating oil. Just a light coat of the slippery stuff will keep dirt and mud from building up on the metal, and your bike ride will be the pleasure you hoped it would be. And once you're home, a quick spray from the garden hose will make the dirt or mud slide right off.

Rust-Bustin' Remedy

1 cup of petroleum jelly
¼ cup of lanolin*

There's nothing worse than heading out to the shed to get your outdoor chores done and discovering that your favorite pair of hedge clippers (or your lawn mower blade) has been taken over by rust. You can prevent this from happening to your metal belongings with this easy fix.

DIRECTIONS: Put the ingredients in a double boiler, and stir over low heat until the mixture is smooth. While it's still warm, pour the solution into a clean jar, apply it to the metal you'd like to protect, and let it dry. Put a lid on the jar and store at room temperature. Then simply reheat the remedy whenever you need to remove more rust.

*Available at drugstores and online.

Dig It 🏠 TO KEEP YOUR SHOVELS, hoes, rakes, and other tools free from rust, make a protective sandpit. Start by adding sand to a 5-gallon pail until it is about two-thirds full, and then mix in ½ quart of motor oil (used is fine). Set the pail in a corner, and after working hard in your yard, plunge your tools up and down in the sand mixture a few times to quickly clean them up. The sand will scrub off dirt and rust, and the oil will give the metal a light coat of protection. You can either leave the tools in the bucket until you need them, or wipe the sand off and hang them back up where they belong.

Super Shortcuts

✂ One of the best ways to keep rust from getting its grip on metal tools is by spraying them with a rust inhibitor. The old-fashioned method is to wipe the metal down with a cloth that's been dabbed with mineral oil. Nowadays, you can do the job in half the time with rust-inhibitor sprays. These products leave a thin waxy coating on the metal. Use them to loosen existing rust so that it rubs off easily, then follow up with a light coat to keep the brown stuff away for good.

Scent-Sational Car Wash

Next time your car could use a bath, why not keep the money you'd spend on an automatic car wash, and do the job yourself? You can give your vehicle a professional-quality wash—with ingredients that smell great—right in your own driveway.

> ¼ cup of liquid hand soap
> 2 drops of orange oil
> 2 gal. of water

DIRECTIONS: Mix all of the ingredients together, then dip a sponge into the wash and soap up your car. Use a garden hose to wash the suds away, rinsing well. Then dry everything off with a soft, clean towel. Your car will sparkle and shine as brightly as it has in years!

Nice Kitty! 🏠 IF YOU LIVE IN SNOW and ice country, keep a bag of clay cat litter in the trunk of your car all winter long. Then if you get stuck, pour the litter under your tires for instant traction. As a plus, the weight of the bag in your trunk will help you control your rear-wheel-drive vehicle better in slippery, snowy weather.

Winter Storm Warning 🏠 IF YOU'RE TIRED OF SCRAPING snow and ice off your car windows, make it easier on yourself with this trick: Use an old rug runner to cover the windshield. Place it carpet side down, and secure it to the window by closing the ends in the front doors. In the morning, you can just roll the rug up and hit the road without lifting a scraper.

That's Brilliant!

When you need to pull off the road at night or in bad weather, make sure you and your car are visible to other drivers. Turn on your hazard lights and set out any reflectors or flares you have (that's why you should always carry them!). If you need to change a flat tire, leave your trunk open so that the trunk light stays on and illuminates the whole back of your car. In short, do whatever you need to do to be seen.

Small-Job Concrete Formula

> 53 lbs. of concrete sand
> 27 lbs. of Type I
> Portland cement
> 55 lbs. of ½-inch gravel
> About 1½ gal. of water

It's said that 1 cubic foot of concrete is the maximum that a person can mix by hand. More than that and you need to rely on a powerful mixer to do the job. For smaller jobs though, try mixing up a batch of this concrete by hand (and give yourself a workout while you're at it).

DIRECTIONS: Pour the sand into a wheelbarrow and spread it out evenly. Dump the cement on top of the sand, and mix it thoroughly. Add the gravel to the mixture, and combine all three together evenly. Make a hole in the middle of the mixture, and gradually pour in the water. Use a hoe to turn it all together toward the center. Once all the ingredients are thoroughly combined, you're ready to pour your concrete. **Note:** Always make sure to wear gloves, goggles, protective clothing, and a dust mask when working with concrete, since the dust can irritate the skin and eyes.

Mix Faster

WANT TO MAKE your concrete-mixing task go a bit faster? Try using wet sand to start—just know that it can't be too wet or too dry. To check its suitability, squeeze a handful of sand into a ball. If it crumbles, then you need to add more water. If water seeps out, then you need to add more sand. It'll be just right when the sandy ball doesn't leave bits of sand clinging to your skin when you pull your hand away. Using wet sand will cut down your mixing time just enough to make the task seem a little less taxing.

Super Shortcuts

Get good results from your concrete by using gravel that's suited to the thickness of the concrete slab you're working toward. The rule of thumb is that the biggest gravel pieces should be no more than one-third the thickness of the slab.

Super Siding Solution

Often, all it takes to spiff up shabby siding is a blast from a pressure washer. If you don't own one of these handy machines, you can always rent one from your local home-improvement center. Then, just spray plain water on the siding, using the lowest pressure setting so you don't wash (or blast) off any of the siding's finish. For really stubborn spots, however, use this super-simple recipe.

> 1 qt. of bleach
> ⅔ cup of trisodium phosphate (TSP)
> ⅓ cup of laundry detergent
> 3 qts. of water

DIRECTIONS: Mix the ingredients together in a bucket. Use the soapy mixture to wash the siding, then immediately rinse it down with a hose. Your siding will sparkle again, and your whole house will look like it's brand new.

Moldy Oldies

ICK! MOLD AND MILDEW are growing all over the siding on your home! The best thing to do is to paint it over, right? Not on your life. Now, you could go to the nearest home-improvement center and buy a mildew killer, but you most certainly already have something far less expensive in your laundry room: bleach. Add about 3 pints of bleach to a bucket filled with 1 gallon of very hot water, and wash down the siding with the solution. You may have to scrub a bit, but that should take care of the foul fungi. Then go ahead and prep the surface to repaint if you want, but I suspect that after this treatment, you won't have to!

Q *My siding looked so grimy that I went at it with hot, soapy water and a scrub brush. Unfortunately, it didn't come clean. Before I start in on round 2, should I be trying a different method?*

A If your siding has gotten really dirty, or if you're prepping it for painting, rent some power-washing gear, and let 'er rip. Try to schedule your cleanup for a weekday—you're likely to get a better deal on the rental price. And if the minimum rental period is a full day and it won't take you that long to finish your house, see if you can find a neighbor who'd like to go halfsies on the time—and the cost.

Surefire Shoe Deodorizer

Have your sneakers seen better days? Before you start shopping for a replacement pair, check the care label to see if it's safe to wash them. If the answer is "yes," grab an old toothbrush and give this easy cleaner a try.

> 1⅓ cups of baking soda
>
> 4 cups of warm water

DIRECTIONS: Mix the baking soda with the water in a large bowl. Use the toothbrush to work the formula into the surface of your shoes, then go over them again to get into all the little crevices. Remove the insoles (if possible), and scrub them, too. Let the shoes stand for a few minutes, then rinse them off well. Repeat the process, if needed, before letting them air-dry.

Shipshape Sneakers ⌂ TO KEEP YOUR freshly cleaned tennies in tip-top shape, leave them out of the dryer and away from the fireplace, heater, or oven. Heat can damage the adhesive that holds the sole in place, or it can shrink your shoes down a size. The best way to dry freshly washed shoes is to blot out as much water as you can with a clean bath towel. Then stuff the shoes with paper towels, and set them outside to air-dry. Replace the paper towels once or twice to help the insides dry faster. And don't ever use newspaper to do this job because the ink will smear, smudge, and/or run—and ruin your shoes!

Odor Eaters ⌂ WHEN YOU'RE WORKING or playing hard in the great outdoors, your footgear can get a tad, well, aromatic. To make those boots or shoes wearable in polite company again, here's all you need to do: For each shoe, pour a few teaspoons of baking soda on a square of cotton cloth. Tie the corners of the cloth together tightly, tuck the ball into the shoe, and let it sit overnight. Come morning, your shoes will be fresh as daisies.

That's Brilliant!

When your canvas shoes get a lot of gunk built up on them, here's an easy way to get them clean again. Just scrub them with a toothbrush dipped in carpet shampoo. Rub the fabric until the shampoo foams up, then rinse with clear water.

Tent-Sensational Mildew Remover

> 1 cup of bottled lemon juice
> 1 cup of salt
> 1 gal. of hot water

Whenever mold and mildew get a grip on your tent, they can cause the urethane coating to peel right off as the foul fungi work their way into the fabric underneath the protective layer. So if you notice a discolored area or telltale cross-shaped marks, follow these steps to stop the nasty stuff in its tracks.

DIRECTIONS: Mix the ingredients together in a bucket, then wash your tent with hot, soapy water as you normally would, and rinse well. While it's still wet, rub any mildewed areas with the lemony mixture. Let the tent dry in the sun, then peel off any bits of loose coating, and apply a commercial water repellent to the damaged areas. If you frequently camp in damp conditions, consider applying a product like Tent Guard with Ultra Fresh™, which is specially designed to thwart mold and mildew.

Magic with Milk Cartons ⌂ THE NEXT FEW TIMES you run out of milk, go ahead and rinse out the cartons. But instead of tossing them in the recycling bin, cut them in half and stow them with your camping gear. Then after pitching your tent at a campsite, place the empty milk carton halves upside down on top of the tent pegs, so you don't trip or bash your toes on the pegs in the dark.

GRANDMA'S OLD-TIME TIPS

Once, when I was headed off on a camping trip with my Boy Scout troop, I couldn't find hide nor hair of my tent pegs. There was no time to buy new ones, but Grandma Putt knew just what to do: She gave me a bunch of wooden clothespins to use instead. And believe it or not, they worked just fine. In fact, I still keep a stash of 'em with my camping supplies for just that purpose!

Terrific Teak Oil Finish

> 1 part linseed oil
> 1 part turpentine

Help your outdoor teak furniture keep its water resistance and bring out its gentle sheen by wiping it down with my homemade potion. It works just as well as the store-bought stuff, and it costs a whole lot less, too! Be sure to wear rubber gloves and work in a well-ventilated area because the noxious turpentine fumes can irritate your nose. And don't mix the stuff or work with it near a fire or a burning cigarette because these oils are highly flammable.

DIRECTIONS: Pour both ingredients into a glass jar with a tight-fitting lid, and shake it well to blend. Put some of the potion on a dry cloth, and wipe the wood in the direction of the grain. Let it sit for at least an hour or more, so the wood can absorb the oil before you use the furniture.

The King of 'Em All

IF YOU'RE LOOKING for outdoor furniture that you can pass on to your children—and their children's children—teak is the biggest bargain you could find in a month of Sundays. If it's well crafted and given proper care—which in this case means simple, periodic cleaning—this stuff will last for centuries. If buying it new is out of the question, look for bargains at auctions and estate sales (especially those in high-end neighborhoods). Trust me, the search will be well worth the effort!

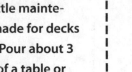

Super Shortcuts

Teak will stand up to decades of wear and tear with very little maintenance. Once a year, coat the legs with a water repellent (the ones made for decks work fine). And here's the simplest way I've found to do this chore: Pour about 3 inches of the water repellent into a clean coffee can. Place one leg of a table or chair into the can, and leave it there for two or three minutes. Then dip a rag into the repellent and coat the leg with it, using an up-and-down motion. Repeat the process until each leg of your deck or patio furniture has had its turn in the can.

Tool-Time Rust-Spot Fixer

Dang! Your favorite shovel picked up a little rust over the winter. Or maybe you found a steal of a deal on some vintage tools at an auction, but they're sporting a few rusty patches. No problem! This fixer will get rid of the rust in no time flat.

> 2 tbsp. of salt
> 1 tbsp. of lemon juice
> Nonstick cooking spray

DIRECTIONS: Mix the salt and lemon juice together, and rub the rusty spots away. Then rinse with clear water and dry thoroughly. To guard against future trouble, spritz the metal surface lightly with the cooking spray.

Nuts and Bolts 🏠 TO LOOSEN A RUSTED screw or bolt, spray it with ammonia. Give the liquid a few minutes to penetrate, and that stubborn piece of hardware should twist right out.

Winterize Your Tools 🏠 AS GARDENING SEASON winds down, but before you put your yard tools away for the winter, protect them from destructive rust by rubbing the metal parts down with a bit of petroleum jelly. That way, they'll be ready and able to do their jobs come spring—without any extra effort from you.

PINCHING PENNIES

One of the fastest ways to remove rust from metal is to pop a wire wheel on the end of an electric drill, and whirl the stuff away. But if you don't have a wire wheel, get the job done with this homemade doohickey that's just as efficient. For each wheel, cut 12 5-inch-diameter circles from a metal screen, and drill a ¼-inch hole in the center of each circle. Slip the stack of discs onto a bolt with a washer on each side, and hold them in place with a nut. Put the end of the bolt into the opening on your drill where a bit would go, and tighten it securely in place. Then start your drill and go to town on your rust removal!

Umbrell-ific Mildew Chaser

Mildew stains are a common problem on umbrellas because of the moisture that lingers in the folds. So when you see any mildew spots, use this quick and easy homemade mixture to make them vanish.

DIRECTIONS: Mix the ingredients in a handheld sprayer bottle. Spritz the stained areas until the fabric is saturated, and let the solution soak for 10 minutes. Scrub the stains vigorously with a cloth, and rinse the umbrella thoroughly. Repeat the treatment if needed. And to prevent mildew from making a comeback, always open the umbrella and let it air-dry after it's been out in the rain.

Umbrella Drip Tray 🏠 WHAT DO YOU GET when you combine an out-of-service umbrella with a crystal chandelier? A match made in heaven—at least that's what Grandma called the combo on cleaning day! She simply opened the umbrella and hooked the handle over the center of the light fixture. Then she mixed 2 teaspoons of rubbing alcohol in 2 cups of warm water in a handheld sprayer bottle, applied the solution to the chandelier, and let the drips fall where they may.

Under the Umbrella 🏠 YOU'LL NEVER GUESS my secret weapon for caring for a vinyl umbrella, but here's a hint: It comes from the automotive supply store! First, clean your umbrella with the cheapest vinyl car top cleaner you can buy, and then protect it against the elements with a protectant specially made for vinyl car tops—again, the least expensive one you can find. Trust me, they work like a charm!

Super Shortcuts

✂ Here's a terrific trick that'll save some scrubbing—take your patio umbrella to a self-serve car wash. Open it up, lay it on its side, and spray it down just as you would your car. Hold the hose at least 2 feet away from the fabric so the force of the water doesn't damage it. Rinse thoroughly, but skip the wax cycle, and then take your nice clean sunshade home to air-dry.

Vivacious Vinyl-Cushion Cleaner

Everyone loves a comfy chair, even when sitting outside. Vinyl cushions are just the ticket for making your outdoor furniture as cozy as can be. And this simple formula is perfect for cleaning outdoor-furniture cushions, your car's upholstery, or anything else that's made of vinyl.

> ½ cup of baby shampoo
> ¼ cup of rubbing alcohol
> 2 cups of water

DIRECTIONS: Mix the ingredients in a bowl or small bucket. Wipe the formula onto the surface with a soft, clean cloth (or an old cotton sock whose mate has gone astray). Then buff with a second cloth. **Note:** This cleaner also works like a dream on leather.

Vinyl Shine ⬆ KEEP YOUR VINYL upholstery looking its best with this three-step routine: First, clean the vinyl coverings with a damp cloth dipped in baking soda. Follow up by washing with a mild solution of dishwashing liquid and water, and rinse thoroughly. Your vinyl will sparkle and shine!

Let's Dish ⬆ DISHWASHING LIQUID is the best all-purpose cleaner for just about anything made of vinyl. It cuts through dirt and grease in a flash, and it won't harm the material. So whenever a kiddie pool, patio furniture, storage container, kitchen floor, or other vinyl surface needs a good scrubbing, squirt some dishwashing liquid in warm water and sponge the grunge away. All you have to do then is hose it down (or wipe it dry, if it's inside) and it'll be ready to use again.

PINCHING PENNIES

Vinyl fences and garden trellises are priced higher than their wooden counterparts, but they require much less maintenance—and time is money! You'll save both when it comes to cleanup. Just use a soapy solution of water and dishwashing liquid along with a soft-bristled brush or cloth to rub the dirt away. Rinse the pieces clean with a garden hose, and you're good to go. Remember to do this work on a cloudy day, so the detergent doesn't leave streaks on your nice clean fence or trellis.

Wake Up and Smell the Coffee Furniture Stain

> 1 cup of ground coffee
>
> 4 cups of water
>
> ½ tsp. of alum*

Wooden furniture looks great in almost any setting, including outside on your deck or patio. But outdoor weather won't be kind to your wooden furniture, so put this stain on your wood to keep nature's wrath from wreaking too much havoc.

DIRECTIONS: Mix the coffee into the water in a large saucepan and simmer on low heat for about an hour, adding more water as it evaporates. Let the mixture cool, strain out the coffee, and mix in the alum as a fixative. Then paint the stain onto the wood surface and let it dry. Repeat the process until you reach your desired shade of brown.

*Available in the spice aisle of your supermarket.

Sweeping Compound 🏠 TO CLEAN UP LOOSE dirt on the patio without raising a cloud of dust, use sweeping compound, which you'll find at most home-improvement and hardware stores. Sprinkle the compound lightly over the patio and use a push broom to whisk away the dust and debris. The fine-screened sand in the compound prevents the loose dirt from going airborne.

GRANDMA'S OLD-TIME TIPS

Outdoor aluminum furniture has come a long way since my Grandma Putt's day. Nowadays, manufacturers coat it with special paint that's designed to help the furniture hold up outside. But whether your aluminum pieces are old or new, bare or coated, you should care for them the same way Grandma did to keep them looking great: Sponge it down with a mixture of warm water and dishwashing liquid. Pay particular attention to the joints and crevices of the frame, and sponge the cushions or webbing with the same mixture to freshen up the seats. Finish by rinsing everything off with a garden hose until all the soap bubbles are gone.

Wet-Step Safety Paint

> 1 can of paint
> ½ cup of clean builder's sand

It's no secret that wet steps are treacherous, whether they're made of wood or concrete. This fabulous fixer gives you a simple way to improve traction year-round.

DIRECTIONS: Mix the sand into the can of paint, and coat the steps with it—it's as easy as that. Just be sure to use the grittiest sand you can find—not the soft, fine-textured kind used for sandboxes and decorative plantings. The rougher the texture is, the better the traction will be.

Slippery When Wet 🏠 WOODEN DECK STEPS get so slippery when it rains that they're a disaster waiting to happen. But it's an easy fix: Just add strips of self-adhesive skid tape (available at home centers) to each step. First, give the steps a good old-fashioned cleaning using the Down-Home Deck and Porch Perfectionizer on page 322. Once the wood is dry, apply the strips to the front edge of each step, following the package instructions.

Soapy Scrub 🏠 DO YOU HAVE FRESH OIL STAINS on your concrete driveway or garage floor? Remove them lickety-split by scrubbing the spots with a mixture of grease-cutting dishwashing liquid and water. Just squirt the dishwashing liquid—no water yet!—directly onto the oil, and wet down the area outside of the stain with water, so that the oil doesn't spread any farther. Use a scrub brush to vigorously work the dishwashing liquid into the stain, add a little water, and scrub to create suds. Then wash the stain away. Now that was easy!

If you have a big oil stain on your concrete driveway, here's a cheap way to clean it up. Cover the oil stain with clay cat litter, and let it sit for at least an hour to soak up all of the grease. Next, apply a concrete degreaser or driveway cleaner, and let it soak in for about 15 minutes. Finally, rinse away the cleaner while you sweep away the litter with a broom. If the oil stains are old and dried on (not fresh spills), skip the cat litter treatment and go straight for the degreaser.

Windshield-Scratch Rubout

Small scratches in your windshield can be down-right distracting—especially when you're driving at night. Light from other cars can reflect in the scrapes, causing an annoying (and possibly dangerous) glare. Use this simple rub to buff the abrasions out so you can see clearly any time of day.

> Electric drill
> Jeweler's rouge*
> 4 oz. of ammonia
> 1 gal. of water

DIRECTIONS: Mount a small lamb's-wool buffing pad on the drill. Rub jeweler's rouge onto the buffing pad until it's coated. Place the pad against the scratched windshield, turn on the drill, and polish the area, working back and forth across the scrapes. Then mix the ammonia and water in a bucket, and sponge the liquid directly onto the windshield. Wash it thoroughly, then dry it with paper towels or a clean chamois.

*Available online and at some jewelry stores.

Swipe the Smoke

IF YOU SMOKE inside your car, then you know how grimy your car windows can get. Make 'em sparkle like new with a mixture of 1 quart of water and 2 tablespoons of lemon juice. Simply wipe it on, and clean it off with an absorbent cloth. The solution will cut through the smoky residue on the windows lickety-split and leave a nice pleasant scent behind.

That's Brilliant!

Good grief. You just got home from the grocery store and noticed a big ol' scratch on your car door. Panic time? Not at all. That is, not if you have some clear nail polish on hand. Clean the scratch with a damp rag and dry it well. Then give it a coat or two of nail polish, which will keep the scratch clean and dry and prevent rust from developing. Just remember that this is a temporary fix. As soon as possible, you (or a pro) should sand, prime, and paint the area with matching touch-up paint from an auto-supply store or car dealer.

Windshield Wonder Fluid

If there's one liquid everyone in the northern states knows you need to buy in bulk during the winter months, it's windshield washer fluid. Washing away salty, dirty, icy spray is practically a full-time, never-ending job. You can save a buck or two next winter by keeping a gallon of this homemade helper on hand.

> 3 ¼ cups of rubbing alcohol
> 2 ½ tsp. of liquid laundry detergent
> Water

DIRECTIONS: Mix the alcohol and detergent in an empty plastic gallon milk jug, then fill the balance with water. Fill your car's windshield washer reservoir with this fantastic fluid, and you'll be all set to hit the road.

I Can See! 🏠 WASH YOUR WINDSHIELD regularly and you'll always see clearly. And if you keep your windshield clean, there's less dirt and grime to collect on the wiper blades, so they'll last longer. To clean the wiper blades, pull each wiper away from the windshield, apply ammonia to a rag or napkin, and rub it down both sides of the blade.

Q *I always wipe down my car windows, but for some reason, they end up looking streaky and smudgy anyway. Is there a trick to keeping windows streak-free?*

FAQ ?

A If you have streaks on your car windows, there are a few tips that you can try. For instance, wash your windows in the shade. Sunlight makes them dry too quickly and causes streaks. When you dry your windows, use a horizontal motion on one side and a vertical motion on the other. That way, you'll know immediately if the streaks are on the inside or outside. And, of course, use a good towel made of an absorbent material to dry your car windows. The best driers to use are made of linen, chamois, or newspaper.

Wow-erful Whitewash

Got an old fence that needs painting? Then channel your inner Tom Sawyer and whip up this whitewash that's guaranteed to brighten up any outdoor structure. When you're finished, tell the neighborhood kids how much fun the task was, and maybe next time, they'll do the work for you!

> 2 gal. of water
> 2 lbs. of table salt
> 6½ lbs. of hydrated masonry lime

DIRECTIONS: Wearing goggles, rubber gloves, and protective clothing, add the water and salt to a 5-gallon bucket, and stir with a paint stick until the salt has dissolved. Then slowly add the lime, stirring until the mixture has the consistency of thin paint. Add more lime or water as necessary. Grab a large paintbrush (or pass a bunch of 'em out to the kids), and have at it!

Reflect Blazin' Rays

HERE'S A NEAT TRICK that'll keep you growing like a pro: Paint whitewash on your greenhouse windows in the spring to help deflect the intense summer sun. That way, your plants will be protected through the growing season. Rain will eventually rinse the whitewash from the windows, leaving the greenhouse exposed to full sun in the late fall and early winter—a time when extra heat and light are more than welcome. Then repeat the application each spring.

GRANDMA'S OLD-TIME TIPS

Grandma Putt taught me all about "whitewashing," which is an old-fashioned gardening remedy that still works today. When painted on the trunks of young, thin-barked trees, whitewash helps combat the insects and diseases that can burrow into the bark and destroy the trees. It also helps prevent sunscald and heat injury in tree bark in the hot, sunny regions of the South and Southwest. So protect your treasured trees with a coat of good ol' whitewash, and they'll reward you with lots of fruit, flowers, and shade!

Yard Tool Handle-Saving Solution

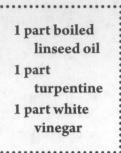

> 1 part boiled
>   linseed oil
> 1 part
>   turpentine
> 1 part white
>   vinegar

Wooden yard tool handles have a tough life, what with working in mud and muck, and sometimes even being left outside in the elements. Use this tonic to keep the wood in fine fettle, so it doesn't crack and splinter. Wipe all of your handles down about once a month, or whenever you notice that the wood is looking a little pale and dry.

DIRECTIONS: Mix the ingredients in a glass jar that's got a tight-fitting lid. Pour some of the mixture on a dry microfiber cloth or old cotton sock, and rub it up and down the handle until the whole thing is covered. Wait about 10 minutes for it to soak in, and then repeat the treatment. Wait another 10 minutes, then wipe any excess off with a clean, dry cloth. Store the leftover potion in a cool, dark place and it'll keep indefinitely.

Splinter Protection 🏠 KEEP THE WOODEN HANDLES of all your tools smooth and splinter-free by applying a thin coat of hair spray. (Either pump or aerosol will do the job.) You'll never have to fear a splinter from handling your trusty trowel again.

Garden-Carrier Lubrication 🏠 LUBRICATE THE WHEELS of garden carts, lawn mowers, and wheelbarrows by smearing petroleum jelly around the cylinders. They will glide across your lawn with ease, no matter how heavy your load.

PINCHING PENNIES

There's no doubt about it: You can get some great bargains on big-ticket power tools at garage and moving sales. But unless the price is so low that you can afford to toss the gizmo out if it breaks down, think twice. Usually, the best place to buy major outdoor power equipment—new or used—is the same place you'd buy a car: from an honest, reliable dealer. You'll get a warranty, and when you need parts or service, you'll know right where to go.

INDEX

feeding, 305, 314, 315
Hit 'Em Hard
Houseplant-Pest
Potion, 307
Houseplant Perk-Up
Potion, 308
light for, 304, 310
pest control for, 308
reducing weight of, 305
Root, Root, Root for
Houseplants Potion,
311
Smooth Leaf Shiner-
Upper, 312
stakes for, 308
Tea-rific Treat, 313
Turbo-Boosted Wonder
Water, 314
watering, 304, 313, 314
Winter Wonder Drug,
315
How Dry It Is Carpet Bath,
161
Hydrogen peroxide
bathroom cleaning, 52,
53, 56, 68
carpet care, 167
clogged drains, 51
floor care, 156
flowerpot cleaning, 310
kitchen cleaning, 4, 33,
37
laundry use, 75, 86
stain removal, 177, 186,
190

I

Ice cubes, 10, 138, 173, 282
Industrial absorbent, 14
Infant Kitten Formula, 239
Ink-Eradicating Paste, 184
Ink stains, 164, 184, 194

Insect bites or stings, 280,
282, 286. *See also* Pest
control; *specific pests*
Inside/Outside Fridge Fixer,
19
Instant-Aging Formula for
Terra-Cotta Pots, 309
Intensive Care Wall
Wipe-Up, 84

J

Jewelry care, 107, 120, 121
Just for Love Dog Delights,
240

K

Kelp, 315
Kerosene, 113, 317, 319
Ketchup, 42
Keys, identifying, 330
Kid-Pleasing Play Clay, 210
Kitchen cabinets
hardware, 25
Kitchen Cupboard
Cleanup Formula, 20
No-Rinse Surface Scrub,
24
Oak-Kay Cabinet
Cleaner, 25
pest control, 278
Surefire Cabinet
Shine-Up Formula, 147
Kitchens. *See also specific
surface or appliance*
All-Around Hard-Hitting
Kitchen Cleaner, 2
deodorizers, 2
Kitchen Grease Buster, 21
pest control, 278, 282
Super-Sanitizing Wipe,
66
Knee pads, 49

L

Laminate countertops, 8, 12,
13, 16, 24
Lamps and lamp shades, 111
Laundry. *See also* Dry
cleaning; Stain removal
care labels, 89, 191
Clothespin Cleaner, 73
Crochet Reviver, 74
Delicate Duds Presoak
Potion, 75
delicate fabrics, 75, 79
Fragile-Fabric Soap, 79
Freshen-Up Fabric Spray,
80
Hair's the Fabric
Softener, 82
Laundry Detergent
Supercharger, 85
mildewed, 83
natural fibers, 74, 77, 80,
92
new clothes, 88, 213
pet messes in, 248
quilts and vintage fabrics,
91
reducing wrinkles, 80
smelly, 76
smoke odors, 86
Smoke-Smell Banishing
Rinse, 89
socks, 82
soured, 75
Super-Softening Laundry
Sachet, 90
synthetic fabrics, 77, 79,
80, 177
during travel, 179
water use, 90
Wool Sweater Wash, 92
Laundry bags, 85
Laundry baskets, 85

Medications, outdated, 58
Melted-Crayon Miracle
    Mixer, 185
Mentholated rub, 290
Meow-velous Move 'Em
    Out Mixer, 243
Mesh produce bags, 42, 48
Metals. *See specific metals*
Mice (pests), 288
Mice (pets), 122
Microwave Gunk Buster, 23
Mighty Mold and Moss
    Remover, 332
Mighty Mouse Control Mix,
    288
Mildew
    All-Surface Cleanup
        Solution, 317
    in bathrooms, 46
    on decks, 332
    on fabrics, 83, 192
    on leather, 146, 192
    Move Out Mildew Spray,
        59
    on siding, 338
    Straight-Shootin' Mold
        and Mildew Killer, 192
    Tent-Sensational Mildew
        Remover, 340
    Umbrell-ific Mildew
        Chaser, 343
Milk
    baths, 218
    floor care, 156
    houseplant care, 312
    for stain removal, 184,
        194
    on wood furniture, 135
Milk cartons, 340
Millet, for birds, 264, 266
Mineral oil
    concrete cleaning, 183

kitchen cleaning, 33, 39
    pest control, 289
    pet care, 220
    slate care, 12, 97
Mineral spirits. *See* Paint
    thinner
Mini blinds, 65
Mint, 200, 274, 276, 288
Mirrors
    Amazing Ammonia
        Cleaner, 41
    Bathroom Shine-Up
        Solution, 42
    hair spray on, 42
    for houseplants, 304
    Magical Mirror Mix, 58
    scratched, 58
    Super-Sanitizing Wipe,
        66
Mockingbird Mania, 261
Modeling clay, 78, 203, 210,
    213
Moist towelettes, 50
Molasses, 331
Mold, 332, 338
Mop buckets, 158
Mosey On Garden Stain
    Mixer, 86
Mosquito bites, 280, 289
Mosquito control, 277, 278,
    289, 291, 292, 294
Moss, 309, 332
Moth control, 81, 278, 283,
    286, 297
Motor oil, for rust preven-
    tion, 335
Mouthwash
    bathroom cleaning, 55
    flowerpot cleaning, 310
    houseplant care, 304, 315
    mildew remedy, 192
    mosquito bites, 289

pet mess deodorizer, 244
Move On, Mosquitoes
    Mixer, 289
Move Out Mildew Spray, 59
Mr. Clean® Magic Eraser®,
    43
Mud, on carpet, 163, 170
Muddy Footprint Fixer, 165
Multipurpose cleaners
    All-Around Hard-Hitting
        Kitchen Cleaner, 2
    All-Around Super
        Cleaner, 153
    All-Surface Cleanup
        Solution, 317
    Easy All-Surface
        Cleaning Elixir, 52
    Everything and the
        Kitchen Sink Cleaner,
        104
    lemon, 60
    pine oil, 8
Murphy® Oil Soap, 25, 151
Mustard stains, 88
Musty odors, 98

N

Nail polish
    First Response for Nail
        Polish Spills, 106
    spills, 52
    uses, 52, 166, 169, 347
Nail polish remover, 156,
    318
Naval jelly, 87
Newspaper, 103, 158
Nijer (Nyjer™) seed, 258
Nix the Nicotine Buildup,
    186
No-Fuss Floor Cleaner, 166
No More Hazy Film
    Formula, 333

Vim and Vinegar Leather
Lotion, 150
Vinegar
baseboard cleaning, 95
bathroom cleaning, 42,
49, 61, 62, 63, 64, 65
car care, 318
carpet care, 165, 168, 244
ceramic items, 35
comb and brush care, 45,
47
cut-flower preservative,
303
as deodorizer, 2, 18, 51,
119, 122
drain care, 17
Easy Herbal Vinegar,
200
fireplace cleaning, 105
floor care, 155, 160, 162,
164, 166, 169, 171, 172,
173, 174, 189
furniture care, 130, 133,
146, 148, 150
hard-water deposits, 67
horse care, 249
jewelry care, 107
kitchen cleaning, 3, 4, 5,
7, 10, 13, 14, 19, 20, 21,
22, 27, 28, 29, 30, 35,
36, 115
laundry use, 82, 83, 86,
87, 88, 89, 182
metal cleaning, 96, 101
pest control, 13, 233,
283, 285
pet messes, 189, 244
skunk odor, 295, 296
stain removal, 72, 176,
178, 187, 189, 191, 193,
194, 195
tool care, 350

vase cleaning, 302
vinyl care, 149
wall washing, 117
window washing, 60,
113, 127, 128, 324
Vintage fabrics, 91, 192
Vinyl floors, 155, 174, 183
Vinyl upholstery, 149, 344
Vitamins, for houseplants,
305, 308
Vivacious Vinyl-Cushion
Cleaner, 344
Vodka, 69, 307
Vomiting, in dogs, 245
Vomit Stain Vanishing Spray,
195

W

Wagons, 108, 334
Wake Up and Smell the
Coffee Furniture Stain,
345
Wake Up and Smell the
Furniture Cleaner!, 151
Wake Up and Smell the
Windows Spray, 128
Wallpaper, 117, 185, 193
Walls
All-Around Hard-Hitting
Kitchen Cleaner, 2
candle wax on, 43
crayon marks on, 43, 185,
188
Daisy-Fresh Spruce-Up
Formula, 49
dusting, 131
Easy Brick Cleaner, 180
fingerprints on, 66
Intensive Care Wall
Wipe-Up, 84
No-Rinse Surface Scrub,
24

painting, 118
Paneling Perker-Upper,
116
pencil marks on, 188
Perfect Paneling Potion,
117
pet urine on, 189
scuffed, 117
Textured-Plaster-Pleasing
Cleaner, 193
Wall-Repairing Plaster
Paste, 129
washing, 129, 131
Wonder-Working Wall
Wash, 131
Washcloths, 226, 246
Washing machines, 83, 89
Washing soda
floor care, 153, 157, 166,
171, 173
laundry use, 85, 182
stain removal, 183
Water gardens, 291
Water marks
on leather, 140
Water-Mark Wonder
Wipe, 130
on wood, 130, 133, 169
Waterproofing agent, 116
Waxed paper, 333, 334
Weekly Wash for Vinyl, 174
Wet-Step Safety Paint, 346
What's Cookin' Oven
Cleaner, 38
Whiteflies, 283
White Line Disease Soak,
249
Whitewash, 349
Wicker furniture, 326
Windex® substitute, 41
Windows
air leaks, 100